Fruit Production and Processing Technology

Fruit Production and Processing Technology

– Author –
Satish Tomar Sinha

Wordpen Academics
Publishing Science That Matter

ISBN: 978-93-90438-53-2 (Hardback)

Published by : **WORDPEN ACADEMICS**
4736/23, Ansari Road, Darya Ganj,
New Delhi – 110 002
Phone: +91-011-35004336
E-mail: info@wordpenacademics.com

Printed at **:** **Replika Press Pvt. Ltd.**

Exclusive Distributor: ***Bio-Green Books, New Delhi***

Preface

Fruits are very important for human beings and also play an important role in religious practices, mythology and art. They are not only delicious but also have many nutrients which are necessary for human health. India is the second largest producer of fruits in the world. Fruit production requires a lot of science and some basic fundamental knowledge to grow them successfully. Today, we have well established world trade networks and sophisticated cultural and postharvest technologies that allow fruits to be enjoyed throughout much of the year, instead of mere weeks per year like our ancestors experienced. Global trade has made formerly rare and exotic treats derived from fruit crops commonplace in countries with no hope of cultivating the plants. Fruit crops are important agricultural commodities, adding tens of billions of dollars per year to the global economy, and being major sources of income for developing countries. Worldwide, over 100 million acres of land has been devoted to their production, and the livelihood of literally millions of farming families depends on continued global trade.

The subject of fruit and nut production deals with intensive culture of perennial plants, the fruits of which have economic significance (a nut is a fruit, botanically). It is one part of the broad subject of horticulture, which also encompasses vegetable growing and production of ornamentals and flowers. This article places further arbitrary limitations in that it does not encompass a number of very important perennial fruit crops covered elsewhere, including vanilla, coffee, and the oil-producing tung tree and oil palm (see coffee, fat and oil processing, wine, and articles on individual plants [e.g., vanilla; tung tree; and oil palm]). Botanists define a fruit in broad terms as the fleshy or dry ripened ovary surrounding the seed of a plant. A pomologist, or specialist in the science and practice of fruit growing, defines it somewhat more narrowly as the fleshy edible part of a perennial plant associated with development of the flower. A nut is any seed or fruit consisting of a kernel, usually oily, surrounded by a hard or brittle shell. Most edible nuts—e.g., almond, walnut, cashew, pecan, pistachio, etc.—are well known as dessert nuts. Not all nuts are edible. Some, used as sources of oil or fat, may be regarded as oil seeds; others are used for ornament. The botanical definition of a nut, based on features of form and structure (morphology), is more restrictive: a hard, dry, one-celled, one-seeded fruit that does not split open at maturity. Among the nuts that fit both the botanical and popular conception are the acorn, chestnut, and filbert; other so-called nuts may be botanically a seed (Brazil nut), a legume (peanut [groundnut]), or a drupe (almond and coconut). In this article the term nut is used in its broadest sense unless otherwise indicated. Improvements in technology and consolidation of the fruit and nut industries in the most favoured climates of the world have been responsible for a steady increase in yield. Thus, the total acreage or number of plants devoted to various fruit and nut crops has dropped, remained about the same, or not risen in proportion to the increase in the respective crop production

This book compiles many fundamental issues of fruit production like layout and planting, many cultural practices, growth and bearing habits of fruit crops, unfruitfulness etc. Understanding of all these topics will help the students for proper knowledge of of fruit production.

Author

Contents

1

Tropical and Subtropical Fruit Crops

Bacterial citrus canker and now citrus greening have ravaged citrus plantings throughout Miami-Dade and Broward Counties and much of south Florida. Infected trees may severely defoliate and management options are very limited; sanitation procedures and following quarantine regulations are very important.

Because of the highly infectious nature of these citrus diseases fruit crops other than citrus should be considered for planting in the home landscape. For some residents, that could mean the opportunity to grow some exciting tropical and subtropical fruit crop alternatives.

The south Florida climate offers the opportunity to successfully produce a surprising variety of tropical and subtropical backyard fruits. Many of them have strange names, odd shapes and exotic flavours, particularly for those of us who have come from up north. Yet many of these fruits are tempting, nutritious, and luscious old favourites enjoyed by the peoples of the tropics around the globe.

Examples include, avocado, mango, guava, and papaya or perhaps the even more exotic carambola, jakfruit and jaboticaba. Although many of these species can attain the size of large trees, routine pruning can keep them to a manageable size for most backyard situations.

The University of Florida, Institute of Food and Agricultural Sciences has a wealth of information on how to grow and enjoy these and many other tropical and subtropical fruits right here in south Florida. Learn which varieties are adapted to our area, recommended cultural practices, pitfalls, and preparation techniques. Below is an overview of some of the fruits we think local residents might want to grow and several (guava and sapodilla) that the University of Florida, IFAS does not recommend because of their invasive status.

Atemoya (custard apple)

The atemoya is a hybrid between the sugar apple and cherimoya. It produces a fruit very similar to the sugar apple, also called sweetsop or anon. It is a small to medium sized, open, deciduous tree with a rounded canopy rarely exceeding 20 feet in height and width.

Because of their relatively small stature, atemoya trees are suited to small yards. A relative of the atemoya, sugar apple has similar characteristics. The fruit is heart-shaped, round, ovate or conical, from 2 to 4 inches in diameter. The pulp is white or creamy white, with a custard-like consistency and a sweet, pleasant flavor.

When ripe, atemoyas become light green or yellow-green in colour. The pulp of atemoya is consumed principally as a dessert fruit and has an excellent flavor. It may be eaten fresh but it may also be used to make a tasty ice cream or milkshake. Season: late Aug.-Oct., sometimes Dec.-Jan.

Avocado

Perhaps one of the best known subtropical fruits, avocado trees are typically medium to large in size (40-60 feet), easy to grow and can produce large quantities of fruit. Limbs are vulnerable to wind damage and can even break under the weight of a heavy fruit load. The fruit is a very large berry consisting of a single large seed surrounded by a buttery pulp.

Mature fruit are generally green, although some cultivars may be black, red or purple when soft (ripe). Avocados do not ripen until they are picked or fall to the ground. Mature fruit size varies considerably depending upon cultivar and growing conditions. Avocados are highly nutritious. Fruit are popular eaten fresh, in salads or used to make guacamole and other dishes. Planting in a well-drained site is a must, as avocados do not tolerate flooding. Productivity, season of maturity, cold tolerance, and disease tolerance vary greatly depending upon the variety under consideration. Season: late May to March.

Banana

A true tropical favourite, bananas are perhaps the best known of the tropical fruits. There are many different cultivars available with a wide variation in fruit type and quality. Bananas are not cold tolerant and some varieties are better adapted to south Florida than others. Growth is extremely rapid during the very warm, wet summer months. Because they reproduce from underground rhizomes and not seed, a single plant can quickly spread out producing multiple trunks in a matter of a few months. Fruits develop in clusters on the end of flower stalks usually within 1-2 years. Shortly after bearing, the stem dies. Bananas like full sun and moist but well drained soil. They will tolerate partial shade, but best growth and fruit production is in full sun. Bananas may be eaten fresh, fried, baked, and added as a component to deserts and drinks. Season: year round.

Caimito (star apple)

Native to Central America and the West Indies, this evergreen tree is highly ornamental and produces a sweet purple or green fruit. Trees may reach 25 to 100 ft. in height and have a dense, broad crown. The attractive leaves are glossy and green on the upper surface and golden brown on the under surface. Trees produce either light green or purplish coloured fruit. The fruit is round, 2-4 inches in diameter, and may have a dark-purple or white flesh. The flesh is milky, sweet, and gelatinous and surrounds up to 10 seeds. Plant trees in a well-drained soil and sunny location. Season: Jan.-June.

Carambola (star fruit)

Originally from southeast Asia, star fruit is becoming increasingly popular and available in the U.S. Locally, commercial production occurs in the southern half of south Florida. Trees are generally small to medium in height (35 feet max.) and spreading with the majority of fruit borne on branches in the mid-canopy region. The fruit is a large fleshy berry, 2-6 inches in length, yellow skinned with a waxy surface and star-shaped cross section. Flesh is light to dark yellow, translucent, crisp and very juicy. Better cultivars have a nice, slightly sweet to subacid flavor. Eat fresh, cut up in fruit salads, or as an iced juice drink. The fruit may also be canned, preserved, and dried.

Carambola trees also have ornamental value with their dark green foliage and attractive flowers and fruit. Plant in a well-drained soil and sunny location, which has protection from wind. On alkaline soils (those with high pH) watch for signs of minor element deficiencies, particularly zinc, iron, and manganese and treat accordingly. Season: July-September, Nov.-Feb.

Canistel (egg fruit)

Native to southern Mexico, Belize, Guatemala, and El Salvador, this evergreen tree is ornamental and produces a sweet yellowish-orange top-shaped fruit. Trees may reach 25 ft. or so in height with leaves mostly grouped at the ends of branches. The attractive lanceolate-shaped leaves are glossy and green. Trees produce 3-5 inch long by 2-3 inch diameter, conical-shaped fruit. The maturing fruit changes from green to light green to golden-yellow or pale orange-yellow at maturity. After picking allow fruit to become soft before eating. Soft fruit will have a yellow-orange flesh, which is smooth with a sweet to musky flavor. Eggfruit is eaten fresh and used in making milk shakes. Plant trees in a well-drained soil and sunny location. Season: Dec.-May.

Jaboticaba

Native to Brazil, jaboticaba is an unusual tree that produces a purple, grape-like berry directly upon the trunk and larger branches either singly or in clusters. Under the skin is a whitish pulp with 1 to 4 seeds. The fruit has a pleasant flavor and taste a bit like grape. They can be eaten fresh or made into jam, jellies or wine. Flowering and fruiting occur periodically throughout the year so multiple crops are produced. Other than its fruit, jaboticaba is known for its beautiful multicolored bark, which gives it value in the landscape. The tree is small, slow growing, and bushy, and seldom exceeds 20 feet in Florida. Flowers are small and white, interesting but rather inconspicuous and borne right on the trunk and larger branches. Jaboticaba is relatively hardy but will not tolerate drought. Plant in full sun in a moist but fairly well drained soil. The tree prefers a slightly acid soil, so some special attention will be required to provide the proper nutrition on our alkaline soils. If provided with an ideal growing environment, jaboticaba can be a relatively low maintenance tree requiring little pruning or spraying for pests. Season: variable.

Jackfruit (jakfruit)

Native to India and Malaysia, this attractive large tree has glossy, dark green leaves and produces a very large, oval shaped, rather unusual looking segmented (spiked) fruit. Typically, fruit are produced in clusters of 2 or more at various height intervals along the trunk. The skin of the fruit must be peeled to reveal the succulent, yellow to orange coloured pulp. Because of sticky latex in the peel, coat your hands with vegetable oil prior to peeling a jackfruit. This will make clean up a lot easier. The flavor is sweet, not unlike that of banana or pineapple, but with a strong, fruity aroma and taste. Fruit may be used fresh, fried green, pickled or roasted (seeds). Jackfruit must be planted in flood free, well-drained soils. In south Florida, jakfruit trees have few serious pest or disease problems. Season: spring-fall (some all year round).

Longan

Indigenous to Mayanmar (Burma), southern China, southwest India, Sri Lanka, and the Indochinese peninsula, longan is large tree (30-40 ft.). Longan is an excellent fresh fruit with a pleasant, unique, sweet flavor. Fruit are relative small (about the size of a typical strawberry), round to oval, and borne in loose clusters. When fruit are ripe, the leathery skin develops an attractive golden brown colour. Longan trees are attractive having a dense, round to upright, symmetrical canopy of dark green foliage. They may grow as high as 40 feet.

Plant in sunny, well-drained sites. Trees may begin to bear within 3-5 years of transplanting. However, unreliable bearing is a major constraint. Season: July-Aug.

Lychee (litchi)

Lychee trees are native to southern China and southeast Asia and may grow to 40 or more feet in height. Lychee is an excellent fresh fruit with a pleasant, sweet flavor. Fruit are a relatively small (about the size of a typical strawberry), round to oval, and borne in loose clusters. When fruit are ripe, the leathery skin develops an attractive pinkish red colour. Lychee trees are attractive having a dense, rounded, symmetrical canopy of dark green foliage. They may grow as high as 40 feet. Plant in sunny, well drained sites, preferably where there is some protection from wind. Trees may begin to bear within 3-5 years of transplanting. However, unreliable bearing is a major constraint. Season: June, early July.

Mamey Sapote

Native to Mexico and the Central American lowlands, mamey sapote trees are large and erect reaching a height of about 40 feet in Florida. Mamey sapote makes an excellent specimen tree in the home landscape with handsome glossy foliage. The fruit has a brown, scurfy peel and the flesh of mature fruit is salmon pink to reddish brown in colour and has a unique, sweet flavor. The tree grows well in a variety of soils, but requires good drainage. Mamey sapote can be eaten fresh, and is also excellent for use in ice cream, sherbets, jellies, and preserves. Milkshakes may also be prepared from the flesh. It requires little care and yields a useful, good tasting fruit. Occasionally, an insect known as the Cuban May beetle may cause some defoliation. Season: Jan.-Sept. (some all year).

Mango

Mango trees are native to southeast Asia and India. Hardly a new comer to Florida, mango trees have been planted and enjoyed for over 100 years in Florida. Mango trees are medium to large trees (up to 100 ft.) and there is a wide variety of cultivars available with varying shapes, colours and maturities.

Most mangoes are ripe when the fruit softens slightly and takes on a yellow to orange or red colour and carries a subtle sweet fragrance. Mangoes are a good source of vitamins A and C. They can be eaten fresh or pureed. Perhaps more than any other tropical fruit, mangoes are readily available at many local nurseries and garden centres. Trees are tough and relatively easy to grow. Most varieties of grafted trees will bear in as little as 3-5 years after transplanting. Season: May-Oct.

Papaya

Indigenous to southern Mexico, Central and South America, papaya plants are relatively short lived (1-3 years) and are easily propagated from seed. Papayas are relatively easy to grow so long as they are sited in full sun and have excellent drainage. Common throughout the tropics, papayas are small to large fruits born on the stem of upright semi-herbaceous trunks. Fruit are sweet, have orange to reddish-salmon coloured flesh and contain numerous small black seeds in the interior cavity. Papaya fruit is typically peeled, sliced and consumed fresh. Papaya ring spot virus can be a problem causing stunting and fruit loss; at present there is no control for this disease. Papaya fruit fly is another problem but can be overcome by placing a 3-5 pound paper bag over developing fruit. Amend the soil with plenty of organic matter and fertilize often. Season: year round.

Passion Fruit

Native to South America, passion fruit is a vigorous vine that produces purple, yellow, or reddish coloured fruit containing seeds surrounded by an orange, sweet, watery pulp. The juice is very aromatic and is commonly used to make juice or punch. Plant vines next to a fence or along a trellis in a well drained soil area with full sun. Vines begin to bear within 3-6 months of planting. Season June-Dec.

Sugar Apple (annon, sweetsop)

Native to tropical America, the sugar apple, is also called sweetsop or anon, has been widely planted in home gardens of south Florida because of its high quality fruit. It is a small, open, deciduous tree with a rounded canopy, rarely exceeding 20 feet in height and width. Because of their small stature, sugar apple trees are suited to small yards. The fruit is heart-shaped, round, ovate or conical, from 2 to 4 inches in diameter. The pulp is white or creamy white, with a custard-like consistency and a sweet, pleasant flavor. When ripe, sugar apples become light green or yellow-green in colour. The sugar apple is consumed principally as a dessert fruit. A relative of the sugar apple called atemoya has similar characteristics to sugar apple. The pulp of sugar apple has an excellent flavor and is usually eaten fresh but it may also be used to make tasty ice cream or milkshakes. Season: July-Sept., Nov.-Jan.

Tamarind

Indigenous to tropical Africa, tamarind is a sturdy, slow growing but large (up to 80 ft.) leguminous tree with rough bark, twisting spreading branches and finely pinnate, evergreen foliage. Flowers are pale yellow with red veins and born in small racemes in early summer. The fruit are brown pods measuring 3-8 inches in length containing large seeds embedded in a sticky brown pulp, which is edible. The pulp may have a sweet to sour molasses-like flavor, can be extracted and used to flavor cooking sauces. It is a primary ingredient in certain brands of steak sauce. Because of the many fissures found on its rough textured bark, tamarind trees provide an excellent substrate for growing orchids. Season: April-June.

Invasive Fruit Crops

Guava

Guava has been assessed by the IFAS Invasive Plants Working Group as invasive and not recommended by IFAS for planting in south Florida; guava may be planted in central Florida but should be managed to prevent escape.

Native to southern Mexico and Central America, guava trees are small, seldom exceeding 20 feet and quite easy to grow. Fruit are produced throughout the year, but the bulk of production in Florida occurs during the summer months. Up to a 25 lbs of fruit per tree can be expected three years after planting, increasing to 100-250 lbs per year when trees attain maturity. Fruit are ripe when flesh softens slightly and turns a light yellow. Depending upon the variety, interior flesh can be white, yellow, pink or red. Cover fruit with a paper bag while small and green to prevent fruit fly damage.

Maturing guava must be harvested quickly as birds and insects will readily attack ripe fruit. Guava is tolerant to pruning, which can hasten flower and fruit development. Popular with Hispanic and Asian Americans, guava may be consumed fresh but more often is made into purees, juices and jellies. The fruit has an exotic flavor and aroma. It is an outstanding source of vitamin C, with some varieties having as much as five times the vitamin C content of fresh orange juice. Season: Aug.-Oct., Feb.-March.

Sapodilla (nispero)

Sapodilla has been assessed by the IFAS Invasive Plants Working Group as invasive in south and central Florida and is not recommended by IFAS for planting.

Sapodilla also called chicozapote, is a medium to large tree with small, dark green, glossy leaves. The fruit is round to oval in shape with a brown peel; inside is a light to dark brown, sweet flesh. There are a number of good cultivars available from local nurseries. Trees are well adapted to south Florida and they are very ornamental. Plant trees in well drained areas away from other trees and structures. Trees usually begin bearing after 2-4 years from planting. Season: Feb.-June.

2

Fruit Wealth of India

India has been exporting fresh fruits since several decades. However, in the more recent past, export promotional activities for fresh fruits have received greater attention. The fresh fruits exports which were only 27 208 tonnes in 1983-84 rose to 55 400 tonnes during 1994-95. Certain other fruits like pomegranate, sapota, banana, litchi, apple and strawberry also entered fresh fruit export market in limited quantities. India not only started exporting grapes but also produced good quality raisins and wines from grapes. During 1994, indigenous production of raisin was about 72 000 tonnes, whereas two factories in Maharashtra (Pune and Nasik) produced about 15 lakh bottles of grape wine.

In the processing sector, there are little over 4150 units registered under the Fruit Products Order (FPO) all over the country. During 1993-94, some 35 proposals for setting up 100% export oriented units were approved. Some of the joint ventures with countries like USA, UK, Germany, Netherlands and Switzerland identified banana and mango puree, fruit juice concentrates and frozen fruits as major export products.

The country is endowed with vast resources and current governmental policy is export friendly with no restrictions on export of horticultural produce. The local needs within the country is expanding at a rapid rate for fresh fruits, while for processed fruit products demand is building, particularly in the urban households. The recent economic reforms have helped in rapid expansion of certain sectors of horticultural industry through private corporate funding. Similarly, planned development in horticulture received major thrust during current 8th Five-Year Plan and several integrated developmental projects are under implementation through various central sector schemes.

Research achievements and transfer of tested technology have helped the country in attaining near self-sufficiency in many horticultural crops. Improvement of export earnings and increased number of turnkey projects with foreign collaboration reflect the growing strength of the horticulture industry, particularly in terms of capability to compete in the global market.

Fruit Wealth

India grows almost all types of fruit crops, ranging from evergreen tropical to temperate deciduous types. For the purpose of this lecture, the major tropical fruits grown in India are briefly described below.

Mango

Mango (Mangifera indica L.) is native of India. Based on the detailed study of the genus Mangifera, Mukherjee (1949, 1985) recorded that although this genus might have originated in the region covering Burma, Siam, Indo-China and Malaya peninsula, the mango itself had its birth in Assam-Burma region. The genetic diversity of mango available in India is very rich and at present more than one thousand vegetatively propagated varieties exist in the country. Clonal selection, selection from chance seedlings and breeding efforts have resulted in identification of elite improved varieties of mango for commercial cultivation in the country. In addition to large genepool of mango (M.indica) available in the country, four other related species viz. M. andamanica, M. khasiana, M. sylvatica and M. camptosperma have been reported from the country (Mukherjee 1985).

Mangifera indica germplasm in India falls under two broad categories:

(a) *Seedling Races, Both Wild and Cultivated Types:* This group consists both of monoembryonic and polyembryonic types. Most of the present day commercial varieties are selections from seedlings. A number of polyembryonic varieties have been reported mostly from the coastal areas.

(b) *Horticultural Races—Clonal Lines:* Distinct clonal differences have been observed in the same variety when grown under different agroclimatic regions and propagated vegetatively from different types of parent materials and on different rootstocks. In a famous north Indian variety 'Dashehari', more than 50 types have been under study at the Central Institute of Sub-tropical Horticulture (CISTH), located in Lucknow, U.P. Improved clones of commercial varieties like Dashehari, Langra and Himsagar have already been made.

Area, production and productivity of fruits during 1994-1995

Sl. No.	*Name of Crop*	*Area (ha)*	*Production (mt)*	*Productivity (mt ha^{-1})*
1.	Apple	210734	1183142	5.61
2.	Banana	444443	13168082	29.63
3.	Citrus	442898	3699044	8.35
4.	Grapes	42139	672893	15.97
5.	Guava	123869	1389962	11.22
6.	Litchi	56538	333247	5.89
7.	Mango	1228341	10992314	8.95
8.	Papaya	60982	1373001	22.51
9.	Pineapple	76412	1055182	14.71
10.	Sapota	38503	495022	12.86
11.	Others (including coconut)	2843199	13573078	—
	Total	5568058	47934967	

Source: NHB Production Yearbook 1997. National Horticulture Board, Ministry of Agriculture, Govt. of India, New Delhi, March 1997.

The genepools of mango available in the country have largely been evaluated and from systematic studies the horticulturally important characters like dwarfness, regularity of bearing, season of cropping, fruit characters (size, colour, quality), and resistance to pests and diseases have been located. The identified donor sources have been employed in breeding programmes, which have resulted in release of a number of mango hybrids in the country. Details of the cultivars identified for various important traits/characters as compiled by Chadha and Yadav (1997).

Cultivars Identified for Desirable Traits

Character	*Variety*
Dwarfness	Creeping, Latra, Jerala Dwarf, Rumani, Neelum (north India), Kalapadi (north India), Manjeera, Mahmooda Vikarabad, Arka Aruna, Amrapali (north India)
Sex ratio	Dashehari, Langra, Neelum, Amrapalli, Alphonso, Rajapuri
Self- incompatibility	Dashehari, Langra, Chausa, Himsagar
Fruit size	Banganpalli, Janardhan Pasand, Kensington, Fazli, Langra, Mulgoa.
Skin colour	Janardhan Pasand, Sindhura, Benazir, Fernandin, Suvarna Rekha, Vanraj, Rosa, Kensington, Tommy Atkins, Sensation, Irwin, Arka Puneet, Nazuk Badan, Raspuri
Pulp content	Langra, Dashehari, Pairi, Arka Aruna, Sindhu, Sahib Pasand, Amrapali, Sunderja, Mallika, Bhutto Bombay, Fazli, Khudadad, Chausa, Sora, Vanraj
TSS content	Sahib Pasand, Doodh Peda, Dashehari, Langra, Arka Puneet, Mallika, Lazzat Baksh, Amrapali, Sakkar China, Gopal Bhog, Rani Pasand, Sunderja
Keeping quality	Alphonso, Totapuri, Sunderja, Dashehari, Banganpalli, Kesar, Rataul, Arka Anmol, Arka Puneet
Regular bearing	Neelum, Kalapadi, Totapuri, Arka Anmol, Kesar
Maturity period	
Early	Mithwa, Zardalu, Lazzat Baksh, Guruvarm, Chandrakaran, Panakalu, Velai Kolumban

Contd...

Character	*Variety*
Late	Kaitki, Bathui, Fazli, Neelum, Arka Anmol, Chausa, Fakhr-us-Samar, Puttu, Sora, Sultan-us-Samar, Azam-us-Samar
Processing	Alphonso, Rani Pasand, Sahib Pasand, Padiri,
(pulp/juice/nectar)	Banganpalli, Totapuri, Hyder Saheb, Pedda Rasam, Sukul, Allumpur Baneshan, Neelum, Chausa, Alif Laila, Prabha Shankar, Kesar, Arka Anmol, Arka Puneet
Canned slices	Dashehari, Baneshan, Gundu, Khader, Langra, Bombay Green, Chausa, Safeda
Pickle and chutney	Amlette, Chandrakaran, Karania, Ram Kela and other mango varieties at unripe stage

Although evaluation reports are available for reaction of cultivars to insects and diseases in fields, no donor sources of resistance have yet been identified.

Through clonal selections, high TSS containing varieties (Sahib Pasand, Lakshman Bhog), regular bearing Dashehari (clone No.53), off-season selection (Niranjan), heavy bearing Langra (about 780 kg/plant), dwarf selections of south Indian varieties Rumani and Totapuri and dwarf polyembryonic types (Manipur I and Manipur II) have been identified as promising ones.

Varietal Diversity

In India, about 30 commercial varieties are grown on large scale and there is a distinct consumer preference of varieties in different regions. Isozyme studies revealed that north Indian varieties differ genetically from south Indian varieties. Some of the prominent commercial varieties in different regions of India are:

(i) Northern India—Dashehari, Langra, Bombay Green, Chausa, Fazli.

(ii) Central India—Langra, Dashehari, Chausa, Safeda and Sunderja.

(iii) Western India—Alphonso, Kesar, Mankurad, Vanraj.

(iv) Southern India—Banganpalli, Totapuri, Neelum, Pairi, Mulgoa, Kalepad, Rumani.

(v) Eastern India—Langra, Kishan Bhog, Zardalu, Fazli, Himsagar, Bombai, Malda.

Some of the quality parameters of a few commercial varieties of India.

Variation in Fruit Quality of Mango

Cultivar	*Fruit weight (g)*	*Pulp (%)*	*TSS (Brix)*
Alphonso	230-290	61.0-69.5	15.0-22.0
Amrapali	190-225	61.4-75.5	18.0-22.0
Bombay Green	190-255	60.5-74.3	15.6-19.0
Bangalora/Totapuri	250-300	66.0-70.0	10.0-13.5
Baneshan	260-370	66.1-79.5	10.5-16.0
Chausa	170-240	61.7-67.1	18.0-16.0
Dashehari	130-155	62.6-70.5	17.5-21.5
Himsagar	130-175	60.2-67.6	12.3-16.5
Kesar	260-300	66.6-69.2	20.0-22.0
Langra	145-235	65.5-69.2	14.0-21.0
Mallika	230-260	69.2-71.0	17.5-19.0
Neelum	235-250	66.0-70.0	11.3-11.5
Sunderja	155-320	61.2-75.0	17.5-19.0

Banana

In India, banana is cultivated in 444.4 thousand ha with a total production of 11.83 million tonnes, contributing 31% of the total fruit production. A total of 565 accessions of banana are maintained in field genebank at the NRC for Banana. The collections continued to be maintained at the Indian Institute of Horticultural Research (IIHR), Bangalore (241), TNAU, Coimbatore (125), Banana Research Station, Kannara (190), Central Horticultural Experiment Station, Hajipur (115), and Gujarat Agricultural University Research Station, Kowur (51). Banana cultivars grown in different regions of the country.

Banana Cultivars Grown in Different Regions in India

State	*Cultivar*
Andhra Pradesh	Dwarf Cavendish (AAA), Robusta (AAA), Amritpani (Rasthali, AAB), Thella Chakkrakeli (AAA), Karpoora Chakrakeli (Poovan AAB)
Assam	Jahaji (AAA), Dwarf Cavendish, Bor-Jahaji (AAA, Robusta), Malbhog (AAB), Chinia (AAB), Manohar (ABB), Kanchkol (AAB), Chini Champa (AB), Bhimkol (AAB)
Bihar	Dwarf Cavendish (AAA), Alpan (AAB), Chini Champa (AB), Malbhog (AAB), Muthia (ABB), Kothia (ABB), Monthan (ABB)
Gujarat	Dwarf Cavendish (AAA), Lacatan (AAA), Harichal (Lokhandi, AAA)
Karnataka	Dwarf Cavendish (AAA), Robusta (AAA), Poovan (AAB), Rasabale (AAB, Rasthali), Hill Banana (AAB), Monthan (ABB), Elakki Bale (AB)
Kerala	Nendran (AAB Plantain), Palayakodan (Rasthali), Dwarf Cavendish (AAA), Robusta (AAA), Monthan (ABB), Red Banana (AAA)
Maharashtra	Basrai (Dwarf Cavendish), Robusta (AAA), Lal Velchi (AAB), Safed Velchi (AB), Safed Velchi (AB), Rajeli (AAB, Nendran), Clones of Basrai
Tamil Nadu	Virupakshi (AAB), CO 1 (AAB), Robusta (AAB), Dwarf Cavendish (AAA), Red Banana (AAA), Poovan (AAB), Rasthali (AAB), Nendran (AAB), Monthan (ABB)
West Bengal & Orissa	Champa (AAB), Morthaban (AAB, Rasthali), Amrit Sagar (AAB), Giant Grover (AAA), Lacatan (AAA), Monthan (ABB)

Papaya

Out of 48 species of the genus Carica, edible fruits are obtained only from Carica papaya. In about 6 species, resistance for different biotic and abiotic stresses have been utilized in crop improvement programmes. Virus resistant C. cauliflora and cold hardy C. candamarcensis and C. pentagona have been collected and extensively used in breeding programme at ICAR institutes.

Systematic germplasm collection and evaluation work was carried out in 1965 at the Regional Station, IARI, Pusa (Bihar). Studies on more than 100 germplasm lines of papaya showed that papaya exhibits variability in respect of plant height (1 to 2.5 m), plant vigour (dwarf to vigorous), plant colour (purple to whitish green), flower type (staminate to pistillate), fruit shape (long to round), fruit size (5 to 30 cm diameter), fruit weight (0.090-13.0 kg), fruit number (10-80), skin colour (golden yellow to dark green), flesh colour (pale yellow to blood red), taste (bitter to very sweet), flavour (objectionable odour when overripe—brown to black). Further progeny row plot trials helped in identifying different desirable parental lines useful for breeding programme. The component characters, viz., leaf length, earliness in fruiting, number of nodes to first flowering, single fruit weight and height of plants were found to be important for expression of genetic diversity.

Varietal Improvement : Till date, 15 high-yielding varieties of papaya have been developed in the country, out of which 9 are widely adopted. A few successful hybrids have been developed and heterosis noticed with a wide range of breeding materials. Some of the improved varieties are given here.

Coorg Honey Dew: It is a selection from the variety 'Honey Dew' and is a popular gynodioecious cultivar which bears oblong-shaped fruits with thick flesh and good flavour. The variety can be maintained pure by growing in isolation.

Pusa Dwarf: It is a delicious variety with dwarf plants and medium-sized (1-2 kg) oval fruits. The plant starts bearing from 25 to 30 cm above ground level and is comparatively drought hardy. This variety is very suitable for high-density planting.

Pusa Giant: Plants are vigorous, sturdy and tolerant to strong wind. It is a dioecious cultivar with big-sized (2.5-3 kg) fruits, suitable for canning industry.

Pusa Majesty: A gynodioecious line, tolerant to viral diseases and root-knot nematodes. The variety is suitable for papain production and is comparable to CO_2 for papain yield. Fruits are medium-sized, round in shape and have good keeping quality.

Pusa Delicious: This is a gynodioecious line with medium-tall plants, and has heavy yield and good qulity fruits (10° - 13° Brix). The fruits are medium-sized with deep orange flesh with excellent flavour.

CO. 1: The variety is evolved by sibmating the type 'Ranchi' over 8 years. It is a dwarf plant, producing the first fruit within 60-75 cm from the ground level. Fruit is medium-sized, spherical, has smooth greenish-yellow skin, nipple slightly ridged at the apex, flesh orange-yellow, soft, firm. It is moderately juicy with good keeping quality. The objectionable papain odour is practically absent in the fruits.

CO. 2: It is a selection purified from a local type at Agricultural College and Research Institute, Coimbatore. Fruits are medium-sized, obovate, greenish yellow, ridged at the apex, flesh red in colour, soft to firm, moderately juicy with good keeping quality. It is a suitable type for extraction of papain. It gives 4-6 g dried papain/fruit.

CO. 3: The fruit of this hybrid (CO.2 x Sun Rise Solo) is larger in size when compared with Solo and exhibits all the desirable attributes of Solo. Total soluble solids (TSS) is as high as 13.8° Brix and average fruits weight ranges from 835 g to 1.5 kg. It has hermaphrodite character and thereby the cost of cultivation is reduced. The sex ratio of this hybrid is 55 female: 45 hermaphrodite flowers. Each tree gives 100 to 120 fruits during the bearing period of 2 years.

CO. 4: This is a hybrid of CO. 1 x Washington. The fruits is medium-sized with round shape similar to Co. 3. Fruit flesh is thick and yellow with purple tinge. Tree has attactive colouration in all parts of the plant and is suitable for home gardens. Mean fruit weight is 1.3 to 12.5 kg. Each tree gives 80 fruits over a period of 2 years. The quality of the fruits is bettter than CO. 1 and Washington, with a TSS content of 13° Brix.

CO. 5: It is a selection from Washington and isolated for its high papain production which has high proteolytic activity and 72.2% protein. It produces consistently 14-15 g dry papain/fruit. It gives 75-80 fruits/tree in 2 years with an average yield of 1100-1600 kg dried papain/ha.

CO. 6: It is a selection from variety Giant. Trees are medium-statured. Fruit shape is round with yellow flesh. Mean fruit weight is 2-3 kg and it is suitable for table purpose as well as for latex extraction, but the quality of latex is relatively poor. It yields 80-100 fruits/tree in 2 years.

At the IIHR, Hessarghatta, 2 gynodioecous hybrids, viz., 39 and 54, developed from crosses of Sunrise Solo x Pink Flesh Sweet and Waimanalo × Pink Flesh Sweet have been found promising with high TSS (14.5° Brix) and better shelf-life. A dwarf mutant Pusa Nanha with prolific fruiting habit has been developed by treating papaya seeds with 15 kr gamma-rays. This variety is most suitable for high density orcharding and for kitchen gardens.

Citrus

Most of the species under the genus Citrus are native to tropical and subtropical regions of Asia, particularly, India, China and in the region between these two countries. The northeastern region of India is considered as one of the natural home of Citrus, at least for a few species. Certain non-edible species like Citrus indica, C. ichangensis, C. macroptera and C. latipes are found growing in wild and semi-wild state in the northeastern region. Bhattacharya and Dutta (1956) considered them as indigenous to the area. Also different strains of citron (C. medica), sour pummelo (C. megalaxycarpa), rough lemon (C. jambhiri) and sour orange (C. aurantium) have been found to grow in northeastern India. It is possible that since very early days, different citrus species might have been distributed in the geographically contiguous regions which are natural home of citrus diversity, particularly between China and India.

Mandarin orange is concentrated in Maharashtra state (especially Nagpur, Amaravati, Wardha and Yevatmal districts) and in the northeastern region of India (states like Assam, Arunachal Pradesh, Meghalaya, etc.). In the Brahmaputra valley of Assam, particularly in the Tinsukia- Dibrugarh region, excellent quality oranges are produced.

Limited area of Karnataka (Coorg), Tamil Nadu (Nilgiri, Palani and Shevory Hills) and Kerala (Wynad) states in the south are also covered under mandarins. Satpura Hills of Madhya Pradesh, adjoining Vidarbha region of Maharashtra also grow good quality mandarins. Of late, Kinnow orange, a hybrid of King x Willow Leaf mandarin has spread to a great extent in the northwestern India, especially in Punjab, parts of Himachal Pradesh, Uttar Pradesh and Rajasthan where it occupies about 27 thousand ha. In Punjab state alone, about 18 000 ha were covered under Kinnow upto 1989 and it is proposed to cover 25 000 ha by 2000 A.D.

In the foothills of Himachal Pradesh, parts of Rajasthan, Maharashtra, Tamil Nadu and Karnataka states also, Kinnow is being grown on commercial scale. Under semi-arid irrigated areas, kinnow comes up well. In Nagpur and Akola in Maharashtra, it performed very well. Kinnow is, however, not successful in tropical parts of south India.

Sweet oranges are adapted well to arid tropics and subtropics. Sweet oranges are commercially grown in Andhra Pradesh, Maharashtra, Punjab, and parts of Tamil Nadu, Rajasthan and Uttar Pradesh. Sweet orange cultivar in Andhra Pradesh is Sathgudi, which is akin to Batavia orange. In western and central India sweet orange cultivar Mosambi is predominant, whereas in the northwestern India, Malta, Jaffa and Valencia are important varieties.

Acid lime is grown on commercial scale in Tamil Nadu, Andhra Pradesh, Maharashtra and Karnataka states. In Andhra Pradesh alone, about 27 000 ha is covered under acid lime. Lemons are grown commercially only on a limited scale. Eureka lemon had been found to be promising but has not spread well. In northeastern India, a lemon-citron type known as 'Assam Lemon' is popular and has found favour in homestead gardens.

Fruit Crops

Passion Fruit

A promising hybrid between purple and yellow-fruited varieties has been evolved at Chettalli. The hybrid is high yielding and tolerant to leaf spot, root rot and resistant to root knot nematodes.

Guava

Two soft seeded guava varieties Arka Mridula and Arka Amulya have been released.

Annona

One interspecific hybrid in annona named as Arka Sahan has been released which has good keeping quality and high TSS.

Litchi

One selection in litchi has been identified and released from CHES, Ranchi under the name Swarna Roopa which is resistant to fruit cracking with high TSS. This is the first litchi variety released in the country.

Jujube/Ber

A promising ber selection CHES-1 from variety Umran has been released which is earlier than Umran, high yielder and escapes insect, pest disease problems and has been christened as Gomati Kirti.

Pomegranate

Two pomegranate hybrid with dark red arils and soft seeded superior to Ganesh seedless have been developed and are in the advanced stages of testing.

Grape

Grape is grown in 42 139 ha with 672.9 thousand tonnes of production. Most of the commercial varieties are introductions from abroad. Two wild species, namely, Vitis lanata and V. riparia are reported from northwestern Himalayan foothills. The commercial varieties are both seeded and seedless types.

Seeded Cultivars

Anab-e-Shahi: The origin of this cultivar seems to be a bud sport as it has a satelite chromosome. It has been one of the most productive cultivar grown in India and yielded over 12-16 tonnes/ha/year. It has attractive large bunch of berries with good shipping quality.

Bangalore Blue: It is reported to be a vinifera × labrusca hybrid. It is medium in vigour and yield. The bunches are small and compact. The berries are small to medium and dark blackish purple in colour. The ripening is uniform. Apart from being used for table purpose, it is being extensively used for juice and wine making. It is known for its hardiness and resistance to disease for which it finds a suitable place as parent in a breeding programme aimed at inducing disease resistance.

Black Champa: It is a selection made at the Indian Institute of Horticultural Research (IIHR), Bangalore. The vines are vigorous with medium yielding capacity. The seeded grape of excellent quality is suitable for table juice and red dessert wine. It is susceptible to cracking and rotting during rains.

Cheema Sahebi: A selection from open pollinated seedlings of "Pandhari Sahebi' in India. The vine is vigorous and a very heavy yielder. Bunches are long, conical and shouldered with medium-sized, oval and pale berries. Shipping quality is poor due to weak pedicellar attachment. It is a late-ripening cultivar.

Gulabi: It is also known as Karachi, Paneer Drakshi and Muscat. It resembles Muscat Hamburg of Australia. The vine is medium in vigour and yield; bunches are small and loose with deep purple, small and spherical berries. It has thick-skinned berries which attribute to good keeping quality. It has a muscat flavour. The ripening is early and fairly uniform.

Seedless Cultivars

Perlette: Berries are medium in size, whitish green and spherical. Flesh is soft and mild muscat flavoured. It has good keeping quality. Small underdeveloped berries (shot berries) scattered all over the bunch is a major defect.

Pusa Seedless: It is a selection made at the Indian Agricultural Research Institute, New Delhi from unknown origin. It resembles Thompson Seedless in many characters. Vine is vigorous and medium yielder. Apart from table purpose, good quality raisin can also be prepared from it. It has good keeping quality.

Thompson Seedless: It is one of the most commercially important variety in India. It is a table and a raisin cultivar.

It has wide adaptability and has performed well in all the grape growing regions of the country. It is mid-season and uniform in ripening.

Vine is medium in vigour, the bunch is medium-large, long, conical to cylindrical, shouldered and compact. The berries are yellowish green to golden yellow when fully ripe, small and elongated. Eating and keeping quality are excellent. Berry size is improved through GA application.

Composition of some underutilized fruits in India

Constituents (%)	*Jujube*	*Aonla*	*Jackfruit*	*Bael*	*Custard apple*
Moisture	*81.0*	*81.2*	*77.2*	*61.5*	*73.6*
TSS	*13-20*	*10-14*	-	-	*26.4*
Acidity	*0.2-0.8*	*1.4-2.5*	-	-	*0.3*
Protein	*0.9-1.7*	*0.5*	*1.9*	*1.8*	*0.8*
Carbohydrates	*12.8*	*14.1*	*18.9*	*31.8*	-
Total sugar	*5.4-12.4*	*4.0-9.8*	*20.6*	-	*21.8*
Vitamins (mg/100g)	*70-165 (Vit. C)*	*600 (Vit. C)*	*540 IU (Vit. A)*	*931 IU (Vit. A)*	*8.5 (Vit. C)*

Arkavati: It is a hybrid developed at IIHR, Bangalore. It is a high yielding variety, suitable for raisin making.

IIHR also released other varieties like Arka Kanchan (seeded), Arka Shyam, Arka Hans, Arka Neelmani (black seedless), Arka Shweta (white seedless). One black hybrid Arka Krishna has been released for juice making, while Arka Soma and Arka Trishna are good for wine preparation.

Underutilized Tropical Fruits

The underutilized fruits like Jujube or ber (Ziziphus mauritiana), aonla or Indian gooseberry (Emblica officinalis), jackfruit (Artocarpus heterophyllus), bael (Aegle marmelos), custard apple (Annona squamosa), jamun (Syzygium cuminii), karonda (Carissa congesta), tamarind (Tamarindus indica) are spread over the tropical and subtropical belts of the country. Some of these have good nutritive value and considerable local demand (Pareek and Sharma 1993).

Ber (Ziziphus mauritiana)

Indian ber or jujube belongs to the genus Ziziphus of the Rhamnaceae family. Out of about 50 species of the genus, about 20 occur in India. It is the hardiest cultivated fruit tree grown in north Indian Plains.

In the arid Rajasthan state, about 750 ha area was under the crop in 1991. Indian ber starts vegetative growth with commencement of rainy season and sheds its leaves during summer. For good cropping, pruning is essential. Presently, about 70 000 ha is covered under ber in the country.

Cultivar Umran has high yield potential with large sized fruits (av. weight 26.8g), whereas cv. Gola, Reshmi, Illaichi have distinct quality attributes. Gola is highly drought hardy. Unlike Chinese cultivars, Indian ber cultivars are not very tolerant to low temperature. Some cultivars, namely, Safed Rohtak, Sanaur 5, Jhajja Selection and Katha Phal showed resistance to powdery mildew, which is the major disease in ber.

Bael (Aegle marmelos)

Bael belongs to family Rutaceae and is an indigenous fruit of India. The trifoliate aromatic leaves are used for sacred offering in Hindu religious functions. It is reported to have medicinal value; the unripe fruit is often prescribed for diarrhoea and dysentry, while ripe fruit is a tonic, laxative and good for heart. Fruit is a rich source of carotene and riboflavin. There was no standard cultivar in bael. In the recent past, some improved cultivars (NB-1, NB-7, NB-9) have been identified. A large genetic diversity still remains unexplored.

Jackfruit (Artocarpus heterophyllus)

Jackfruit grows wild in the forests of Western and Eastern Ghats of India and a few species, namely, A.chaplasha, A.hirsutus and A.lakoocha are available in nature in Assam, West Bengal in the Eastern India and in the Andaman Islands. Being a cross pollinated plant, propagated mainly by seeds, a wide range of variability is noticed in Jackfruit. There are two broad groups, soft flesh and firm flesh types. The soft flesh types are quite juicy and pulp is highly scented. So far there is no well defined variety, and in different localities, local varieties are known differently. Local selections have been named as Gulabi (rose scented), Champa (flavour like that of Michelia sp.), Hazari (bearing large number of fruits in a tree).

As a result of local survey, some better types have been collected. Since in India raw jackfruits have good demand as vegetable for culinary purpose, emphasis is also given on fruit characters like thinner rind and soft flesh at premature stage of fruit development.

A few selections, namely, NJT_1, NJT_2, NJT_3 and NJT_4 with large fruits and excellent pulp quality have been identified for table purpose, while types like NJC_1, NHC_2, NJC_3 and NJC_4 were found to be better for culinary purpose. North-Eastern region of India produces large quantity of Jackfruit. The state Assam of in eastern India is reported to have about 8000 ha under Jackfruit, while Tamil Nadu state in south India has a covered area of about 2960 ha under this crop.

Jamun (Syzygium cuminii)

Jamun belongs to the family Myrtaceae and is a native of India, Burma (Myanmar) and Sri Lanka. The tall tree with evergreen foliage is an excellent roadside tree and very often used as windbreaker. It is widely grown both in north and south India. Distribution of some economic species of Syzygium. Small dark purple coloured fruits with sub-acid spicy flavour are eaten fresh.

There is no standard variety in jamun. In north India, a type known as 'Ra Jamun' with big (2-5 cm long fruits) fruits is normally grown. A small (1.5-2.0 cm long and 1-1.5 cm diameter) fruited type is also grown for late harvesting. A wide range of variability exists in jamun and survey conducted (Keskar et al. 1989) in Pune and Ahmednagar districts in Maharashtra state revealed wide variation in fruit weight (3.5 to 16.5 g), pulp contents (54-85 %), TSS (4.5-17 %) and acidity (0.16 to 0.55 %). Some promising lines have been selected.

***Syzygium* species of horticultural interest found in India**

Species	*Distribution*
S. cuminii	Indo-Gangetic plains of north India, Tamil Nadu in south, widely distributed
S. arnottianum	Western Ghats, the Nilgiris, Palni and Anamalai Hills
S. bracteatum	Western Ghats, Eastern/northeastern India
S. operculatum	Grows wild in Nilgiri Hills of Tamil Nadu, Western Ghats
S. aqueum	Mainly in Assam, Sikkim and Meghalaya, Eastern/ northeastern India
S. fruiticosum	Grows as an avenue tree, widely distributed

Other Fruits

India grows many other tropical fruit crops. Fruits like guava, litchi, sapota, pomegranate and pineapple are grown commercially. Good germplasm exists in these crops and improved cultivars have been released. In custard apple, a few promising cultivars have been identified. At the IIHR, large number of hybrids obtained from different crosses are being evaluated.

In passion fruit, a cultivar named, 'Kaveri' with multiple resistance against Fusarium sp. leaf spot disease and nematodes has been developed from crosses of Passiflora edulis × P. edulis f. flavicarpa. In sapota, Kalipatti, Cricket Ball and Kirtibarthi are some popular cultivars, while in guava, Allahabad Sufeda and L-49 are popular. In pomegranate, very fast expansion has taken place and some excellent, high yielding cultivars like 'Ganesh' G-137 have been released. In a traditional crop like tamarind, several high yielding types have been selected. Tamil Nadu state alone covers about 13 820 ha under tamarind and produces about 42 000 tonnes of fruit.

Conclusions

The fruit industry in India has made remarkable progress during the last three decades. The area under fruits has increased from 1.22 million ha in 1961 to 5.56 million ha in

1994-95. As per FAO Yearbook 1994, India ranked second after China with a production of about 33.23 million tonnes. Our share in the world production of fruits is about 8% and India produces about 65% of the world mangoes and 11% of the world bananas, ranking first in both the crops.

Although banana occupies less than 12% of the area under fruits, it contributes nearly 32% of the total fruit production in the country. Quantity-wise, out of all fruits produced in the country, maximum production is in banana (32%). The five fruits, mango, banana, citrus, guava and apple, cover about 75% of the total fruits produced in the country. In less known fruits like sapota, jujube/ber, aonla and pomegranate, area expansion and production rise during the recent years is simply spectacular.

3

Genetic Resources and Breeding Patterns

Almost 400 species of edible fruits and nuts are identified and used either commonly or in a limited way in different countries of South-East Asia. Nearly 90% of them are woody plants including 250 tree species that mostly produce fleshy fruits, many of them with recalcitrant seeds. Among the tree species, only about 35% are cultivated in home gardens, around the villages, orchards or plantations and the remaining 65% are in the wild. Nevertheless, the fruits from the forest trees are regularly collected, sold and used.

The commonly cultivated species are about 20 in number. Fruits play an important role in the Asian society especially among the rural people. Besides meeting dietary needs, fruits are used for traditional and cultural uses like offering in temples, ceremonies, giving presents, for new years greetings, weddings, decorations and others. Many species of fruits, wild or cultivated, have medicinal value (Burkill 1966; Verheij and Coronel 1992).

Tree growth in home gardens protects the environment and improves microclimate. In spite of many varieties of fruits said to be available in the tropics, the supply of good fruits is always short and beyond the economic means of majority of the people in rural areas. Besides many species of fruit trees are either irregular or limited in their bearing and fruits available for a short period. Limited quantity or availability always increases market prices. Unlike well known cultivated fruits of temperate countries, the grower can hardly influence the yield level or the period or the duration of fruiting in tropical fruit trees. Many of them are aseasonal in the cycle of reproduction. Some comparative data have been collected in the past in different countries of South-East Asia, of the commonly cultivated species, which show wide variations confirming fluctuations in yield. It is not easy to bring many tree species under cultivation because of the lack of appropriate knowledge, huge investments involved and unpredictable profits.

Growth phenology, floral biology, life cycle, metabolism and function, cytology and embryology, seed production, viability, and seed germination of tropical fruit trees are hardly studied which are important to analyze the causes of fruit production and yield levels that influence the market supplies and the prices. The success of tropical fruit tree cultivation achieved so far is mostly due to traditional knowledge and selection of good varieties. Except for identification, science has contributed very little for fruit tree improvement including its genetics, breeding and yield. Year to year or season to season fluctuations under given conditions need to be carefully monitored to predict good harvest and to understand the fruiting season.

The variations in climatic conditions influence the fruit production. The yield level is more secure and predictable with herbaceous and shrubby species than the trees. How much energy and biomass is apportioned for flower and fruit production among fruit trees is yet to be properly understood. Tree breeding is a time-consuming exercise since tree growth involves many years before they start flowering and to check on first or second generation hybrids. Very few exceptions to all the above are mango, citrus and a few others since they are cultivated for several millennia (Anon. 1996). Annuals are better utilized than perennials but very few herbs produce good table fruits of quality and taste.

Most of the trees are propagated either by seeds or grafts or cuttings of one kind or the other which enhance or limit the genetic variation and breeding patterns. Natural sexual reproduction methods of fruit trees display wide variations. Within the framework of the limitations mentioned above, the germplasm collections, their maintenance and the breeding methods of five fruit tree species, namely, Averrhoa carambola (carambola), Durio zibethinus (durian), Garcinia mangostana (mangosteen), Lansium domesticum (duku/langsat) and Nephelium lappaceum (rambutan) are discussed in this paper. All of them are important cultivated fruit tree species in humid tropical countries of South-East Asia with significant economic value and annual turnover.

Origin and Distribution

Averrhoa carambola (canambola) of Oxalidaceae is said to originate from South-East Asia as well as tropical America. The exact location of the place of origin is unknown. However, this species has been introduced to many countries both in tropics as well as subtropics of old and new worlds. Durio zibethinus (durian) of Bombacaceae is native of South-East Asia and has many closely related species. Altogether, there are 27 species and from Malaysia 23 species of this genus are reported, six of which are edible and only Durio zibethinus is commonly cultivated. This species is also widely distributed in Sumatra, Borneo, Papua New Guinea and other neighbouring countries. Some natural populations are still present in the wild and recently it has been introduced to Australia and other neighbouring countries.

Garcinia mangostana (mangosteen) of Guttiferae is an allotetraploid hybrid with the parentage of G. hombroniana and G. malaccensis. It originated in West Malaysia where both parents are present in the wild. It is cultivated in many countries both in the new and old world. Lansium domesticum of Meliaceae, commonly known as duku/langsat originated in western part of South-East Asia. This species is less well known when compared with other fruit tree species of the region but cultivated in many countries of South-East Asia extending to Australia, Surinam, some Pacific island countries and Puerto Rico.

Naphelium lappaceum (Rambutan) of Sapindaceae is well known in South-East Asia. The origin of this species is attributed to many countries in South and South-East Asia. It is a popular fruit species of the region and commonly cultivated in many parts of South-East Asia.

Diversity in Species and Cultivars

Due to their long history and domestication, many cultivars and varieties of each of the above species have been identified and recommended for cultivation. The cultivars are known either with common names in local language or they are given specific numbers. Three clones of A. carambola are identified in Malaysia as B_2, B_{10} and B_{17}. All of them have sweet nutritious and non-astringent fruits. B_{17} is said to produce better quality fruits than B_2 and B_{10}.

Altogether 19 cultivars of Durian are recognized in Indonesia, Malaysia and Thailand. Chanee and Mathong and Kaan Yao are widely cultivated in Thailand. Five varieties are popular (Petruk, Sunan, Sukan, Sitokong and Simas) in Indonesia. D_{24}, MDUR88, MDUR78 and MDUR79 are the popular varieties in Malaysia. Number of fruits produced per bunch (3-5) is better in Malaysian and Thai varieties than those in Indonesia (1-2).

But the number of fruits produced per tree per year is almost double in Indonesian varieties than those in Malaysia and Thailand. Fruit size plays an important role and mediates the number of total yield. But there is plenty of scope for further improvement. In Brunei, Durio zibethinus is not very common. Fruits of three other species of Durio are commonly collected and sold. One natural hybrid (D. graveolens × D. zibethinus) with few varieties is also popular. All these would indicate that the germplasm of durian is rich and can be made use of to produce new selections/ hybrids with better quality trees that are disease-resistant and more productive.

Cultivars in mangosteen are very few identified by variable characters like fruit shape, colour of the pericarp, etc. For all practical purposes, mangosteen has only one clonal variety since fruits are produced parthenocarpically and seedlings produced remain true to mother type.

About 20 clones of rambutan are under cultivation in different countries of South-East Asia but they all originate from six varieties of N. lappaceum (var. lappaceum, pallens, xanthoides (restricted to Borneo), cuspidatum, eriopetalum and robustum). Recommended clones in Malaysia are R3, R134, R162 and R170; in Indonesia, Rapiah, Tankuweh and Binjai, in Thailand Rognrian and Seechompoo. The clones presently used in Philippines were introduced from Indonesia and they are very popular (Yacoob and Subhadrabandhu 1995).

More varieties of Lansium domesticum have been collected at Sri Wijaya University, Palembang, Sumatra. These are all intermediate or close relatives of the two major varieties of duku and langsat. All of them are referred by local names-Rasuan, 1105, 1089, 1092, Komering 3, 12, 17, Palembang 175, 176, 358, 561, Lonjang - 2, Bulat and Penyandigan. About a dozen more varieties are in the hold, but yet to be characterized. The good varieties in Java are Duku condek and Duku metesih. Their relationship to Sumatran varieties are yet to be ascertained.

Duku and langsat are the two varieties generally recognized in Malaysia and Thailand. Many sub-varieties carry local names like Duku Johore and Duku Trengganu, etc. One seedless variety of Thailand called Longkong Norathiwat, is very popular and it is propagated by cuttings, grafting, etc. (Yacoob and Subhadrabandhu 1995). This is the best variety because of bigger size, thin skin, is easily peelable, sweet and seedless. Paete, Mindanao and Duku are the three varieties cultivated in Philippines, mostly introduced from Indonesia and Malaysia (Coronel 1983). Some of the main points on origin, distribution, habit and ecology are summarized.

Germplasm Collections

South-East Asia is one of the richest regions for plant biodiversity in the world. Many parts of the region are yet to be properly explored to identify the plant resources. Attention was called in 1970s to the urgent need of collecting data on tropical fruit germplasm in South-East Asia and other crop species, and the need for extensive survey of fruit resources in the region was emphasized. Five countries in the region were surveyed and the data on the degree of genetic erosion of fruit tree species were compiled (Sastrpradja 1975). Information on 105 species under 29 genera of 17 families was collected, classifying the wild and cultivated species and the relative status of genetic erosion of each species. Fruit tree germplasm resources available in Malaysia and the Philippines were also reported. 39 species of major fruits, 29 of minor fruits and 49 of rare fruits were recorded for Malaysia (Loh 1975).

Seasonal fruit species were cultivated on 40 000 ha of land and the non seasonal ones on 42 000 ha in Malaysia. Germplasm resources available for breeding in the Philippines included 440 accessions of plants belonging to 36 families (Valmayor and Espino 1975). The varieties present under each species were also systematically listed. The relative quality of each variety and the endemic status of many were recorded. This was the beginning of analysing the fruit germplasm in South-East Asia.

An expert consultation meeting was held in 1994 to discuss tropical fruit species of Asia (Arora and Rao 1995). Eight activities were identified to promote conservation and use of tropical fruit species, which included:

1. Assessment of priority genepools on major and minor fruit species,
2. Gather information on diversity available,
3. Information on ethnobotanical aspects,
4. Use of local knowledge,
5. Classification of genetic diversity,
6. Develop for complementary conservation strategies,
7. In vitro conservation, and
8. Ecogeographic surveys and in situ conservation.

Recommended list of major fruits suggested for different Asian regions included:

South Asia—Mango, jackfruit, citrus, litchi.

South-East Asia—Mango, rambutan, citrus, durian, jackfruit, mangosteen.

East Asia—Litchi, citrus, mango, longan.

It was agreed to give attention to carambola also. As part of documentation needs, it was decided to get the reports prepared from different countries on the current status of genetic resources of various fruits mentioned above. Most of these status reports have been finalized and circulated in APO among the concerned national programmes.

Number of germplasm collection in Thailand for durian, mangosteen and duku are large and most of them are not characterized and classified. Because of the export potential, Thailand grows many varieties of these fruits but the genetic diversity of the germplasm collections of different varieties is yet to be determined. Some papers are published on quality, standardization and group characterization of Thai durian and rambutan (Yacoob and Subhadrabandhu 1995).

The germplasm collection of rambutan in Malaysia is large even though only 8 varieties are recommended for cultivation. The fruit characteristics are well analysed and 193 clones are so far registered. Early bearing hybrids are recognized by their fruit characters. The registered clones are recommended for planting. Hybridization programme is aimed to obtain dwarf varieties that produce quality fruits in great numbers. Attempts are also made to develop varieties that are resistant to pests and diseases. Plans are made to obtain intraspecific hybrids (Rukayah Aman and Salma Idris 1996).

New information is provided regarding germplasm collections of duku/langsat (Lansium domesticum). Germplasm collections are maintained in Padang, Lempang and Palembang, all in Sumatra, Indonesia. IPGRI is supporting a research project carried out in Sri Wijaya University, Palembang. Sixteen varieties have been identified and six of them are characterized. The criteria followed for selection are fruit colour, thickness of pericarp, number of seeds produced per fruit and few vegetative characters like leaf size, shape, petiole length and others. The cytological studies have yet to be done. Germination studies and seedling characters are being studied.

The duku collections in Malaysia are maintained by MARDI and in a few other small gardens. The varieties are yet to be properly characterized. As part of IBPGR collection activity, some germplasm was collected nearly 15 years ago and it is maintained at Songkla University in Thailand. Three varieties are recognized and they need to be properly characterized.

Reproduction and Breeding

Averrhoa spp.

Averrhoa has only two species, A. bilimbi (bilimbi) and A. carambola (carambola). Heterostyly is the common feature in the family Oxalidaceae with distyly or very rarely with tristylous forms. In carambola, flowers with long styles are self-fertile and those with short styles receive pollen from flowers with long style. Occasionally the plants produce only long or short styled flowers, or sometimes both the types of flower are produced on the same plant.

More research is needed to understand the segregation and the pattern of flower biology. No hybridization work has been done. Attempts are made to cross the two species *i.e.* A. bilimbi and A. carambola with no success. Flowers are produced in great numbers and less than 1% of flowers set fruits. The fruit size varies even on a single tree. Much improvement has been done for selection of good varieties and their cultivation. Size and taste of fruits are the two criteria followed to select the good varieties. Seeds are used for propagation.

Durio Zibethinus (durian)

Flowers are produced in clusters and in great numbers. Anthesis is in the late afternoon or early evening. Flowers last for 12-18 hours and pollination is effected by bats and bees late in the evening or early hours of the morning. Less than 1% of flowers develop into fruits. Flowers are self incompatible and cross pollination by hand has aided better fruit set. Cleistogamy is ruled out but bud pollination (pollinating unopened flowers before anthesis) has resulted in better fruit setting than by pollinating the fully opened flowers (Coronel 1983).

The species is heterozygous. No hybridization has been done so for but it is suggested that the hybridization can be relatively easy since durian species in the wild are not that different from the cultivated varieties. Clonal selection technique has helped to identify genetic variability and the selection of improved cultivars. By selection process, even the size of cultivated trees have been moderated to 15-20 m instead of 20-40 m. The branches are close together and canopy is compact in the selected cultivars. Very little work has been done on the reproductive biology of plants but good varieties that produce quality fruits are well selected and cultivated. There is no comparative study on the various varieties to match them with the agroecological conditions.

The differences noted between the varieties are not very significant and it is possible that a sub-variety of one country is identified as a major variety in another country. Collecting and growing all the varieties together in a specific site or center will help to identify and select superior plants even from among the existing well established varieties. Seeds are rarely used for propagation. Grafting is more commonly practised. Hypocotyl is fleshy with abundant cambial activity and shoot tips of superior trees are grafted to hypocotyls. Under in vitro conditions, shoots developed from seedling nodes and they were rooted giving rise to plantlets. Somatic embryogenesis was also established (Goh 1997).

Gracinia Mangostana (mangosteen)

It is a very popular fruit of the tropics. Genus Garcinia has about 300-350 species. Details of 24 species that occur in India, Andaman, Nicobar Island, Myanmar, Thailand, Malaysia, Philippines and Indonesia are well described. All the species of Garcinia are useful in one way or the other since they yield valuable timber, medicine, resin oil, latex (pigment), fodder and edible fruits. Most of them have sour fruits, except for mangos-teen. The sour fruits replace tamarind in the local cooking and acidity is reduced after boiling. Many believe that G. mangostana is a hybrid between G. hambroniana and G. malaccensis. The trees of the former are very much like mangosteen and still present in the wild, whereas trees of G. malaccensis are comparatively rare to find.

Details recorded by all the earlier workers suggest that mangosteen trees that are presently cultivated may represent a single clone of female trees that produce fruits parthenocarpically. The embryo is undifferentiated, seed size varies even within a single fruit, 1-3 carpels produce seeds and the others remain abortive in a five carpelled ovary. The relative role of integuments and nucellus that produce the edible flesh is yet to be well defined. Details of seed development that have been described so far raises more questions and doubts than providing the proper development details. Only two or three good papers are published on the subject and the critical points of development still remain very vague.

G. mangostana is self fertile but male sterile. Since it has many related species which are fully fertile, hybridization is possible, especially with closely related species like G. hambroniana. Intensive, well planned studies will yield good results. Another possibility is to use the tissue culture methods since tissue regeneration and differentiation in vitro is very easy to accomplish in this species. Even the embryonal tissues display great potential for proliferation and mass propagation. Cytologically, the species is very heterogenous indicating the existence of several viable but highly evolved populations. They need to be properly analyzed to unravel the genetic complexity of the species. Populations of mangosteen in different countries are not comparatively well studied. Such studies will help to understand the biodiversity of the species further substantiating the clues provided by varied diploid chromosome numbers. Multidisciplinary research will provide the key for hybridization and to improve the quality of fruits of mangosteen as well as the yield.

True to type seedlings can be obtained since undifferentiated seeds are formed in mangosteen by maternal tissues. Furthermore, each seed has the potential to give rise to 4-8 seedlings when the seed is cut transversely and the segments used. Whole seeds are also used in nurseries and more than one seedling/per seed is obtained. Very interesting results have been obtained with regard to seed size, percentage germination and seedling survival (Coronel 1983). Leaf, stem and cotyledon tissues both from juvenile and mature plants proliferated and gave rise to shoots in vitro. On separation, they could be rooted well, producing plantlets in large numbers. Both axillary and adventitious bud formation was common. The plantlets produced in vitro maintain clonal purity since no abnormal tissues or polyploid tissues were involved (Rao et al. 1982; Rao and Lee 1986; Goh et al. 1988, 1990; Goh 1997).

Lansium Domesticum (duku/langsat)

It is a polymorphic species. The reproductive and vegetative characters are very similar but two or three distinct fruit types are present. The fruits of different varieties vary considerably in colour, texture, shape, taste, number of seeds produced, etc. This species is male sterile since pollen development is irregular. Polyspores (clusters of abortive pollen) are formed instead of regular pollen. Parthenocarpic fruits develop and each fruit will have 1-3 seeds. One particular variety in Southern Thailand produces no seeds at all. The plants are propagated by using seeds, when they are available and by vegetative methods like budding, grafting and others when there are no seeds. L. domesticum has many varieties and some are characterized. Large populations in different countries need to be well analysed to select superior plants. Commercial possibilities are many but presently the fruits are used mostly locally and to some extent in the region.

Since no pollen is produced, the flowers are male sterile but self fertile at least in the two major varieties. The stigma is receptive and it is possible to start hybridization trials by using the pollen of other species of Lansium. At present, there is no data available with regard to three other species of this genus.

Like mangosteen, the vegetative and reproductive tissues of this species respond positively under in vitro conditions showing proliferation, differentiation and organogenesis. But manipulating the tissue growth or by inducing somatic embryogenesis, it is possible to improve the quality of plants and fruits. In both varieties of Lansium domesticum, namely, langsat and duku, enhanced axillary branching was induced under in vitro from nodal explants of seedlings as well as mature trees (Tham 1997). Nephelium lappaceum (rambutan)

It is the only species more commonly cultivated than others. Nephelium has 30 species, very evenly distributed in Indonesia, Malaysia, Thailand and Philippines. More genetic diversity is present in the first two countries and the collection of germplasm is not complete. Even those varieties collected in 1981-82 or 1986 are not property maintained due to lack of adequate funds. Field germplasm maintenance is expensive and labour intensive. Maximum diversity is present in northern and central Borneo. It is not uncommon to find a series of intermediate forms between the wild, semi-wild (small fruits, sour) and cultivated varieties and the well recognized clones under cultivation (big fruits, firm crunchy flesh and sweet) which indicates the fact that large genepool that is present is yet to be properly improved by further hybridization. Once cultivated for a few years and later abandoned, trees revert back to wild state showing many intermediate stages between the wild and cultivated types. Fruit size, taste, good peeling and non-stringy nature of aril are the important criteria to distinguish the superior varieties.

There is no breeding work done so far in rambutan. Flower biology is studied by one or two persons and the results are inconclusive. The sexuality of flowers is interesting since a single tree produces the functionally male or female flowers or occasionally both the types of flowers on the same tree. Natural barriers created to avoid self pollination are many including the phenomenon of protandry and protogyny in a modified form.

The variation in the length of style and spreading of stigma controls the self pollination process in many instances. Large number of flowers are produced in an inflorescence and less than 2% of flowers are pollinated and only 0.3-0.2% of the pollinated flowers develop into fruits (Coronel 1983). It is very common to see one small abortive fruit attached to a well developed one suggesting that one fruit perhaps develops at the expense of the other in terms of energy transfer and material accumulation. Up to end of one or two weeks after pollination, both the fruits develop simultaneously but later, one stops growing while the other continues to develop to maturity. All these details would indicate that intensive studies are required to improve our knowledge on hybridization and fruit production in rambutan. Seeds are still used for propagation in many countries. Budding is more common in many nurseries. Tissue culture protocol is available to raise plantlets using vegetative tissues of mature plants. The production of good fruits of selected varieties can be maintained by planting the in vitro raised plants on a large scale.

Outlook for the Future

Only recently rambutan was classified under major tropical fruits of the South-East Asian region. Durian, mangosteen, duku and star fruit is grouped under minor fruits (Arora and Rao 1995). The relative ranking of each and the extent of cultivation varies from country to country. However, the five fruits discussed presently are not usually consumed or traded outside the region but they have a big commercial potential. Only recently, carambola produced in Malaysia is sold regularly in Holland because of increased and regular supply as well as establishment of dependable marketing channels. Good production strategies are yet to be established for the five species discussed.

The cultivation and improvement of a fruit tree species passes through several stages, namely: (a) collecting from wild or the forest (b) domesticating and growing them in a limited way around homes or gardens and selecting the better varieties or clones (c) establishing small orchards after recognizing their economic potential, and (d) finally cultivating the select and superior varieties on large scale in a plantation with good investment and for handsome economic benefits.

The five fruit tree species discussed in this paper have not reached the plantation levels as banana, citrus or mango have. Most of them are limited to semi-wild conditions or home gardens or small orchard systems. Number of plants cultivated per hectare are small in case of durian and mangosteen, slightly better in case of rambutan and star fruit in that order. Both the area and number of plants cultivated there involves considerable manpower, investments and marketing strategies.

The volume of the total annual production of durian, mangosteen, duku, rambutan and carambola in the region is less than 50% of the total banana production which reflects the limited level of cultivation of the five species. Many improvements need to be made for further advancement. Research and development activities in the following areas are of utmost importance and to be pursued on a priority basis to improve the fruit tree species. A good knowledge of reproductive biology, phenology, vegetative propagation including tissue culture and breeding methods should be well studied. Matching the selected varieties with well identified agroecological conditions would improve the yield. By nature, most of the plants of the five species are well adapted to moist soils, good rainfall and high humid conditions.

The extent of genepool available, and germplasm collected and stored for each of the five species is variable. Selection done to establish cultivars in duku is still at preliminary stage. The collections made for durian, rambutan and mangosteen are fairly extensive. For star fruit, there is limited scope. However, the number of species or varieties used under each genus is very limited and offers good opportunity for further expansion.

Clonal propagation methods used presently are to be streamlined for all of them which would reduce the juvenile period and hasten the onset of reproductive phase. Precise details need to be recorded in each case. Flowering pattern and period of different species vary locally and in the region. Both the early and late flowering varieties need to be identified, selected and properly cultivated to prolong the fruiting season and availability. The plants that flower regularly and abundantly in a population deserve more attention to enrich the germplasm collection.

Position of inflorescences and number of fruits produced per inflorescence varies among the five species. Durian and duku are cauliflorous with limited fruit production compared with the general biomass of the tree. Though with axillary inflorescences, mangosteen is shy bearer and produces small number of fruits. Fruits of rambutan and star fruit are produced in relative large numbers. There is plenty of scope for research to increase fruit production in each of the species.

Sexuality is an enigma in both rambutan and duku. Anthesis and the correct stage of pollination and pollinators involved are to be precisely determined in each case. Fruit production method is asexual and abnormal in case of mangosteen and duku. Many gaps need to be filled for proper understanding of their reproduction.

Tree breeding is a long process in case of tropical fruit trees but a beginning has to be made to achieve success. Pioneer work already done on oranges or apples should set the guideline to plan further research. The limited manpower and finance available in the region set limitations to conduct research and in the development of facilities. Part of the economic gains presently accrued in fruit industry should be set apart for research and development activities.

Lastly, it is important to remember that most of the fruits of these five species are sold in village markets, weekly markets or in roadside stalls, the common shopping areas of the poor people in rural areas. Any improvement planned or extended should be for the benefit of rural communities because they have been the custodians of the germplasm for long time which the researchers plan to use presently to improve the quality of trees and the quantity of fruits produced therefrom. The indigenous knowledge should be tapped to plan further activities for improvement of less well known tropical fruit tree species.

Strategies for Collecting

It is generally recognised that the plant genetic resources (PGR) are the basis for sustained plant improvement. Hence, collecting and conservation of PGR have assumed special significance in the last few years. Collecting plant germplasm can include various aspects, such as survey, exploration, rescue missions, bioprospecting, network activities, etc. Due to charges in the international scene in the recent years, for example Agenda 21, the Convention on Biological Diversity (CBD) and Global Plan of Action (GPA), exploration for monitoring and PGR collecting are expected to increase. Although general principles of collecting plant diversity are well reviewed (Guarino et al. 1995), it may be opportune time to look at closely from point of view of fruit species germplasm collecting.

In the following account, effort is made to present a preliminary strategy for tropical fruit species germplasm collecting (woody and/or perennial). I wish to thank Dr. George Ayad, who has generously shared his ideas on the strategy for coconut germplasm collecting, on which I have drawn much in preparing this lecture. It is essential to remember that we are dealing with variable species group.

Defining the most appropriate strategy for such variable group is a difficult task. The difficulty mainly arises, among others, due to the fact that we have very little information on genetic structure of the populations, which can often be very complex (Allard 1970; Brown and Marshall 1995). We also have insufficient information on the types, amounts and distribution of genetic diversity within and between populations. We may not also be able to pinpoint the imminent threat of genetic erosion facing many of target populations, a factor that determines the urgency for collecting.

The objective of PGR collecting may be achieved if maximum genetic diversity is identified and collected, properly documented and follow up research is undertaken on the collections and its results are made available to the scientific community. However, in perennial species like fruits that we are dealing here, we cannot really afford to collect, at random, that gives us maximum genetic diversity (Ramanatha Rao et al. 1996). We will try and find answers to this as well as many other questions as we go along. However, I need to point out that what will follow is a generalised model that gives the components of a strategy but not the actual strategy. From following discussions, it may be possible to draw up a strategy once a target genepool is identified.

Strategy for Germplasm Collecting

There is a need, initially in very clear terms, to understand the objectives of collecting or the collection itself. The reasons for collecting, for example, are: (1) danger to genetic erosion, (2) clear need for use in improvement, (3) missing in the extant collections, and (4) need for biological studies (Guarino et al. 1995). It is also essential to remember that collecting is not the end of the story.

Setting clear objectives initially will greatly help in defining a strategy (or strategies) for collecting. The objective of any collecting strategy is to develop a pattern of collecting which will provide maximum genetic diversity occurring between and within populations in different or within the same geographic areas. At the same time, it is important in the case of species groups like fruits to be able to collect the most useful material. This is mainly because of their perenniality as well as the long gestation period needed for the plants to become useful and longer time scale needed for any improvement programme. This would entail some level of selective sampling while collecting.

Since there are definite limits to the number of samples that can be collected and handled effectively, the strategy is "to capture the maximum amount of useful genetic diversity with a minimum number of samples" (Marshall and Brown 1983). Once this basic tenet is agreed, then it will be easier to formulate a strategy.

Needs of Users

Given the need for prioritisation of any collecting of agricultural biodiversity, needs of the genebank curators, researchers and breeders assume paramount importance. In such situations, selective or targeted sampling strategy should be adopted. Biased sampling of particular fruit trees in pre-determined sites may have to be carried out.

Types of Collecting Missions

For tropical fruit species, like many other crop/domesticated species, exploration, survey and collecting missions may take the following forms.

Conservation-oriented Missions

In this case, the conservation of genetic diversity in a given genepool per se is the overall objective. For this purpose, random sampling is employed and usually aim is sampling an alleles class termed "locally" distributed (in one or a few adjacent populations) / common (alleles with a frequency greater than 0.05). About 50 - 100 plants per site are sampled with sites representing environmental heterogeneity as much as possible as defined in Marshal and Brown (1975). A point to note here is that sampling 50 to 100 plants may be very difficult in the case of fruit species.

Targeted Collecting

In this type of collecting, plants/populations are targeted for particular traits, *i.e.*, collecting in specific agroecological areas/sites known to or thought of possessing such traits including unique variants (typical collecting strategy of breeders). This presupposes that PGR were collected from these areas earlier and that they have been screened thoroughly for traits under consideration.

Collecting to Mitigate Genetic Erosion

This type, also referred to as emergency collecting, is typical "rescue missions" in areas characterized by high genetic erosion risk or even threatened with extinction due to impending developmental activities. This is also typical collecting for conservation. This availability of information on types and rates of progress of genetic erosion is a basic prerequisite for this type of collecting strategy. An example for this type of collecting could be one that is undertaken in an area that is expected to be submerged due to the construction of a dam.

Output of Collecting Missions

While developing a strategy, along with objectives, the outputs from a mission should be clearly thought out. Some of the outputs may be clearly an offshoot from the actions taken, but it is essential to note them so that any modifications to the strategy can be carried out.

The major output is a set of samples that encompass a "representative" range of "the maximum" genetic diversity in the target species/populations. Along with such samples, information on the samples and their habit becomes available. Since any collecting mission subsumes survey and monitoring aspects, another output may be methodologies to identify species, areas, sites and levels or targets of diversity for collecting and recommending areas for in situ conservation. For example, agroecological information, indicators to erosion, diversity analyses that are done either at pre-collecting stage to select sites for sampling or carried out during the mission, indigenous knowledge (IK), etc., help us to determine such areas which could become part of a protected area system or part of community conservation programme.

Skilled and well trained explorers and collectors of genetic diversity is an offshoot of any mission. Lack of sufficient skilled personnel in PGR activities including germplasm collecting is a major problem for efficient conservation of PGR, and collecting activities help to develop practical skills in this area.

The Target Diversity

For enhancing the efficiency of collecting missions, collectors need to distinguish between various types and structural levels of variation. Some examples are:

- Heritable vs. non-heritable variation.
- Cryptic vs. expressed variability; what collectors see by the eye is "phenotypic variation".
- Morphoagronomic and morphological vs. physiological and biochemical variability.
- Variation between and within populations.
- Ecotypes and agroecotypes vs. geographic races.
- Polymorphism (allelic series for a variety of traits, mean and total number of alleles per locus) and polymorphic loci.
- Simple genetic variation (alleles, heterozygotes but no allelic series), simple loci.
- Selectively neutral vs. adaptive allelic variation.
- Genotypes vs. alleles (objective of conservation: single alleles, gene complexes or super genes, genotypes).
- Measures of diversity, *e.g.* allele and genotype frequencies, gene diversity measure (Nei's gene diversity index), heterozygosity measurement (average heterozygosity/ locus, disequilibrium coefficients.
- Allele distribution, "wide-spread" -occurring in many populations in the area, "local" - occurring in one or a few adjacent populations in the target area.
- Allele frequency. "common" - frequency above 0.05, 'rare' frequency less than 0.05. The common-widespread and

the rare-widespread alleles are independent of the sampling strategy while rare-local alleles is impractical. The focus is on the locally-common alleles. The objective of random sampling strategies (Marshall and Brown 1975) "is to capture at least one copy of 95% of the "locally common" alleles occurring at frequencies higher than 0.05". This will be assured if a sample of 59 random unrelated gametes from the population is made. In sexually reproducing outbreeding species (*e.g.* most of the fruit tree species), this would mean a reference sample size (bench mark criterion) of 30 randomly chosen individuals from a heterogeneous outcrossing population or 59 random individuals in a selfing species.

Essential Pre-requisites

In order to develop a most appropriate collecting strategy for any given genepool and plan efficiently collecting missions, biological and agroecological information need to be gathered and analysed. Without such knowledge, any strategy is doomed to fail. It is not possible to deal with all the information needed, much of it can be found in Guarino et al. (1995). Some examples of such pre-requisites are given here.

Taxonomic Information

As one can see, this is fundamental to any collecting. Phylogenetic information that can provide species relationships, however, basic it might be, can help greatly in collecting germplasm on genepool basis.

The Breeding System

Information is needed on the nature of breeding system of the target species. Information on the levels of outcrossing in the crop species, the amount of gene flow between populations or species, inter- and intra-specific classification of identifiable genotypes, geographical distribution and ecological amplitude, etc.

Information on Past Collecting

A given area might have been explored or even sampled in the past. Any information such as description of regions/

sites sampled, observations on the variation in the field/ habitat, recommendations for further collecting, observations on rate and causes of genetic erosion, characterization and evaluation data, distribution of genetic variation within and between populations from results of analyses of collections, the useful genetic diversity identified and used in crop development programmes, etc., can guide the development of a strategy.

Site Information

Ecogeographic information such as vegetation and soil maps, herbaria records, etc., can provide a full picture on species distribution range and on its environmental heterogeneity.

Indigenous Knowledge

Information about cultivation and agronomic practices, farmers preferences and selection, use of crop cultivars, etc., is the basic information on the incipient genetic diversity of the material targeted.

Decision-making Process

While developing a collecting strategy for a genepool, a number of decisions need to be made about many issues. These decisions are mainly made based on the pre-requisite information that we have just seen. So the accuracy of the information on hand will determine the efficacy of any strategy developed. Some of the decisions that need to be made are discussed below. The information that is used for decision making are mainly genetic, the theoretical basis of which is covered by Dr. Nissila.

As representation of genetic diversity being the major concern in most of the missions, collecting genetic resources should be based on population not on single tree/ plant. However, in the case parennial fruit tree species, considerations such as the gestation period of germplasm resources required to maintain a truly population sample will have to be looked into.

Sampling strategy needs to be determined, *i.e.*, should it be random or non-random (selective) sampling. The objective in conservation of PGR is to maintain 'a representative sample' of the genetic variation contained within or between populations.

This objective can only be met through random sampling. However, some level of selective sample will be required to enhance the value of germplasm to be collected for immediate utilization.

Well ahead of any mission, decisions have to be made about the components of sampling strategies. These include decisions regarding the number and distribution of sites per region, number of plants or trees per site, number of seeds or other propagules per tree or plant to sample, and have to be made based on the biological information as well as on practical feasibility.

There is need to assess the implications of the choice of a strategy on various practical issues such as resources required, time, etc. For example, a critical consideration is that, if the decision is to collect samples from at least 50 plants/site depending on the breeding system from as many sites as possible (depending on the heterogeneity of the environment in the target area), what would this entail in terms of resources required to cover a given area in a certain amount of time.

We also need to determine the size of sampling sites and select sampling sites. The sizes are determined in turn by the size of farmers' field or in terms of a colony of species. The size of a colony may be determined by factors of environmental heterogeneity, by observed variation, etc. Selection of sampling sites depends on pattern of distribution and density of individuals, occurrence of unique variants etc. within a site or colony.

In outbreeders, one has to sample larger number of plants/site and lesser number of populations in a region since the spatial scale of local differentiation in outbreeding populations is likely to be larger than that in autogamous species. In inbreeders (in apomicts), populations diverge for the alleles they contain and for the amount of genetic polymorphism. Hence, sampling greater number of populations within regions even at the expense of the number of plants/site. Increase in sample size of exceptionally highly polymorphic populations may be an optimum strategy. In natural selfing populations, local subpopulation structure can develop in response to microsite/environmental heterogeneity. In such situations "stratified local sampling" at the given site(s) will be appropriate (Brown and Marshall 1995).

Some variations, based on specific needs, have to be made. Special considerations, while collecting wild species, may be required, for example, sampling at evenly-dispersed sites within the target area. In the case of cultivated species, when it is very difficult to determine the samples to collect based on phenotypic variation, clustered sampling of sites within the target area will be useful. As would be the case in collecting most fruit tree and forestry tree species, budwood cuttings of 10-15 individuals per site in close clusters, followed by reduction of sample size/ elimination of samples in the light of progeny tests or other procedures will be extremely useful.

Generally, information on the distribution of genetic diversity between and within populations is either lacking or incomplete. Also, intensive sampling is undertaken following an initial preliminary collecting in the first year and then search for specific genotypes or genes of particular traits, *e.g.*, drought tolerance, disease resistance, etc. In such situations, use of coarse grid sampling will be useful to sample maximum useful diversity. This involves random sampling at wide intervals in an area where the size of the intervals are determined by environmental heterogeneity factors, *e.g.*, climate, soil, vegetation, farming practices, crop cultivars, altitude, latitude, etc.

Conclusions

Some sort of stepwise decision making process needs to be developed so that a collecting strategy for a genepool of fruit species can be developed. Based on the discussions presented so far, which are far from being exhaustive, the following steps are suggested. It should be noted that this is not a blue print for germplasm collecting, but only a conceptual base for a strategy.

Clear and well defined objective(s): Generalised objective is to capture the maximum amount of useful genetic diversity with a minimum number of samples. Prioritise the materials to be targeted, based mainly on the needs for improvement and potential for use.

Determine type of collecting mission based on the objective, however, given the current climate, a combination of conservation, use and genetic erosion concerns need to be addressed. Information on these aspects need to be assembled to make this decision:

- Chalk out the outputs from a collecting mission clearly to facilitate their achievement from the mission.
- Familiarise with various types and structural levels of variation that can be present in target population.
- Based on the taxonomic information, determine the target species (of a genepool under consideration) to be collected.
- Based on information on past collecting, conservation and evaluation activities in the region/ on the genepool, further refine the target area / genepool.
- Based on the objective(s), site information, information on breeding systems and indigenous knowledge, etc., determine sampling techniques that need to be used.
- Sampling strategy needs to be determined based on a number of issues (as described in the section above), which should be basically random sampling with appropriate use of selective and stratified sampling as required.

4

Assessing Diversity Using Molecular Techniques

Plant genetic resources are one of the most valuable assets available to mankind. Protection and conservation of these resources for future generations, therefore, assumes great significance. An important component for effective and efficient management of plant genetic resources as well as their utilization is characterization of the germplasm. Such a characterization is essential not only for identification of various species but also to determine genetic relatedness among them. The information generated could be used successfully in breeding programmes wherever possible. This also assumes great relevance in the present context of intellectual property rights and trade agreements.

Assessment of diversity has traditionally been through morphological characters, which has often found to be rather less effective. Biochemical markers like isozymes were the next to be used and have been the choice in several cases (Scandalios 1969). However, isozyme analysis has its inherent disadvantages like limited number of enzyme loci, and developmental and seasonal dependent enzyme expression. With the advent of molecular biology techniques, DNA-based markers have replaced enzyme markers in germplasm identification and characterization.

Because of its plasticity, ubiquity and stability, DNA is the ideal molecule for such analysis. Although molecular markers have been used to create extensive linkage maps for many annual plants (Helentjaris 1987; Bernatzky and Tanksley 1986; McCouch et al. 1988), not many attempts have been made for their use in fruit crops. Genetic analysis and breeding of woody perennial fruit species can be complicated by many factors including long periods of juvenility, high ploidy levels, lack of described Mendelian markers and so on (Rowland and Levi 1994). Hence, the task of developing molecular marker-based genetic maps is both challenging and important.

Apart from isozymes, a number of DNA-based marker systems are now available for germplasm evaluation and classification. The basic premise is that variation in the nucleotide sequence of DNA can be exploited to produce characteristic fingerprints. The systems available are:

- Restriction fragment length polymorphism (RFLP)
- DNA amplification fingerprinting (DAF which comprises AFLP- amplified fragment length polymorphism / RAPD- random amplified polymorphic DNA)
- Microsatellites
- Cleaved amplified polymorphic sequences (CAPS).

It is proposed to lay emphasis in this paper on the first two DNA-based marker systems along with isozyme-based marker system.

Methodology

Isozyme Analysis : Proteins are attractive for direct genetic studies because they are the primary products of structural genes. Changes in coding base sequence will, under many circumstances, result in corresponding changes in the primary structure of proteins. Single amino acid substitutions can have marked effects on the migration of proteins under an electric field as in electrophoresis.

When this occurs, isozymes are obtained which are different forms of an enzyme exhibiting the same catalytic activity but differ in charge and hence mobility. In isozyme analysis, crude plant extracts are subjected to gel electrophoresis on starch or polyacrylamide gels. Following electrophoresis, the enzymes of interest are detected by probing the gels with specific activity stains. Variation in banding patterns obtained between individual samples can be used to sort out genetically the different varieties tested.

Restriction Fragment L:ength Polymorphism (RFLP)

The detection of a polymorphic locus characterized by a number of variable-length restriction fragments, termed restriction fragment length polymorphisms (RFLPs) was originally used to fingerprint genomes. In this method, DNA is digested with restriction endonucleases which recognize and act on specific sequences in the DNA. The cleaved DNA is subjected to electrophoresis on agarose gels to separate the fragments according to their size.

The separated fragments are transferred to a membrane support and hybridized to labelled DNA probes. Fingerprints revealed as variant banding patterns on X-ray films (or on the membrane itself when non-radioactivity is resorted to) are as a result of the polymorphic fragments detected by the chosen probe. The number of fragments obtained and their sizes would depend upon the number of restriction sites and their positions in the DNA which, in turn, would be characteristic of the genome and the genomic sequence being analyzed. Illustrates a simplified example of RFLP analysis applied to a phylogenetic problem (from G. Kochert in Introduction to RFLP mapping and Plant Breeding Applications, Department of Botany, University of Georgia, Athens, GA).

DNA Amplification Fingerprinting (DAF)

The discovery of the polymerase chain reaction (PCR) method of DNA amplification (Saiki et al. 1988) is an important milestone in molecular biology research. In the polymerase chain reaction, the target DNA is amplified many million-fold using two oligonucleotide primers of about 20 nucleotides in length, that specifically hybridize to opposite DNA strands flanking a region to be amplified.

The reaction is carried out with a thermostable polymerase in the presence of deoxynucleotides and primers in a thermocycler. The reaction mixture is alternately heated to high temperatures and cooled several times. A typical cycle would involve denaturation of DNA and separation of the strands by heating, annealing of the primers by cooling following extension of the primers by DNA polymerase by addition of nucleotides to them. The polymerase chain reaction is thus a cyclic process; with each cycle, the number of DNA targets doubles.

This method has provided geneticists with new ways of generating molecular markers for mapping and genome characterization. One of these is the use of short 10-base primers to randomly amplify polymorphic DNA sequences (RAPDs) (Williams et al. 1990). After the polymerase chain reaction cycles, the amplified products are separated by electrophoresis either on agarose gels and detected with ethidium bromide or on polyacrylamide gels and detected by silver staining. This strategy is called DAF, DNA amplification fingerprinting. DAF uses low stringency amplification conditions so that primers can anneal arbitrarily at multiple sites on each template strand. Although initiation of DNA synthesis occurs throughout the template, only those sequences in which priming sites are on opposite strands and in near proximity will be successfully amplified.

DAF allows easy detection of DNA polymorphisms that can be called amplification-fragment- length polymorphisms (AFLPs) following the use of acroynm RFLP in conventional fingerprinting. The use of the term AFLP over the term RAPD proposed by Williams et al. (1990) is favoured because amplification in DAF is arbitrary but not random. RAPD is also a registered trademark (Caetano-Anolls et al. 1991).

Several mechanisms can account for the variation in length or number of amplification products observed in DAF fingerprints. AFLPs can arise from nucleotide substitutions that create or abolish primer sites, but also from either the deletion, insertion or inversion of a priming site or of a segment between priming sites (Caetano-Anolls et al. 1991). Two related genomes are amplified with DNA polymerase and an arbitrary oligonucleotide primer (arrow) and the resulting amplification products compared. DNA amplification of genome A and B produces three and two DNA fragments respectively. In genome B, an insertion between priming sites produces a larger fragment a, while a deletion decreases the size of fragment b. Additionally, insertion of a large segment of DNA between priming sites (indicated by upward and downward arrows) eliminates fragment c. Several other alternatives are thus possible for the generation of polymorphisms.

Molecular characterization based on RAPD markers is quickly becoming one of the most popular methods because it is technically simpler than using RFLPs. A number of steps can be done away with such as preparation of genomic or cDNA clones, restriction enzyme digestions of DNA, Southern blotting or hybridization with radioactively - labelled probes. As RAPD analyses require much smaller amounts of DNA and less pure DNA than RFLP analyses, the RAPD method is particularly attractive to geneticists working with species recalcitrant to nucleic acid extraction such as woody plant species (Rowland and Levi 1994).

Applications-Case Studies

As mentioned earlier, while molecular markers have been used extensively in annuals, only a few reports are available in the case of woody perennials like fruit trees (Heemstra et al. 1991, Durham et al. 1992; Jarrell et al. 1992; Schnell and Knight 1993; Rowland and Levi 1994; Bhat and Jarret 1995; Schnell et al. 1995; Graham et al. 1996; Herrero et al. 1996; Liou et al. 1996). In this paper, the applications of marker systems in assessing diversity in two important fruit crops viz. mango and citrus will be discussed. A representative case study of each of the commonly used marker systems has been considered.

Isozyme Markers

Isozyme markers have been used to study genetic diversity in the orange subfamily, Aurantiodiae (Herrero et al. 1996). Despite the great economic importance of citrus, its phylogeny and taxonomy remain a matter of controversy. Measuring genetic variability and studying its pattern of distribution are central steps to establish genetic relationships. They are also essential tools in germplasm characterization and conservation to effectively control genetic erosion, to design sampling strategies or core collections and to establish breeding programmes. In order to determine the intraspecific and intrageneric variability within Citrus and Citrus-related species, Herrero et al. (1996) used ten enzymatic systems. These were phosphoglucoisomerase (PGI), phosphoglucomutase (PGM), 6-phosphogluconate dehydrogenase (6PG), aconitase (ACO), malic acid dehydrogenase (MDH), glutamate oxaloacetate transaminase (GOT), superoxide dismutase (SOD), peroxidase (PRX), isocitrate dehydrogenase (IDH) and leucine aminopeptidase (LAP).

Mature leaves from circular twigs were used to obtain crude extracts for electrophoresis following the method of Asins et al.(1995).

Four enzymatic patterns (ACO2, 6PG, MDH and PRXa) could not be interpreted genetically. Allozymes that could be interpreted genetically were Idh, Pgi-2, Lap 8, Pgm-1.8, Got-1.13, Got-2.16 and Mn-Sod 6. The isozyme analysis was able to distinguish all the species and all but one artificial hybrid. Based on isozyme studies, the species with the lowest genotypic variability are Citrus myrtifolia, C. deliciosa, C. paradisi, C. limon and C. sinensis, while Severinia buxifolia shows the highest value. A broad spectrum of heterozygosity values was found in the collection. Lemons, limes and C.bergamia show a very high percentage of heterozygosity which indicates an origin through intraspecific hybridization. It has thus been shown that the orange subfamily presents a great deal of genetic diversity even at the level of isozymes.

RFLP Markers

Characterization of the Citrus genome through analysis of restriction fragment length polymorphisms (RFLPs) has been carried out (Liou et al. 1996). Citrus species are diploids with relatively few chromosomes and small genomes *e.g.* 1 C = 0.38-0.62 pg (Arumuganathan and Earle 1991). The studies described in the paper of Liou et al. (1996) were done with the major objective of examining the nature of RFLPs observed and thus to begin characterization of the Citrus genome. Citrus reticulata Blanco cv. Clementine, C. paradisi Macf. cv Duncan and an F_1 hybrid (LB-1-21) were used to determine what probe-enzyme combinations revealed polymorphisms in Southern analysis. A backcross family (LB-1-21 × 'Clementine') of 65 randomly selected hybrid seedlings was used for same analysis.

The clones used were genomic clones derived from a Pst I genomic library and cDNA clones derived from a cDNA library prepared from mRNA isolated from cold acclimated trifoliate orange. The restriction enzymes used were EcoR I, EcoR v, Bgl II, Hind III, BamH I and Pst I. 73% of the genomic clones examined appeared to detect single-copy sequence's based on RFLP banding patterns, while cDNA clones revealed a lower percentage of single copy sequences. It was shown that the polymorphisms in banding patterns observed primarily resulted from insertions and/or deletions rather than base substitutions.

RAPD Markers

The use of RAPD markers to establish genetic relationships among Mangifera spp. in identification of Mangifera indica L. cultivars and in validation of genetic relationships among them have been demonstrated by Schnell and Knight (1993) and Schnell et al. (1995).

In the first study, RAPD markers were used to estimate genetic relationships among 9 species of Mangifera with the primary objective to determine which species are more closely related to M. indica. Ten arbitrary 10-mer primers were selected from 90 initially tried, based on the simple, consistent and repeatable banding patterns obtained. PCR was carried out using standard procedures and banding patterns were obtained. Data were scored as the presence or absence of a band. Within each species, only bands appearing in more than one genotype were considered for between-species analysis. A pairwise difference matrix between species was determined using the similarity for interval data and the average taxonomic distance. Cluster analysis was performed using the unweighted pair group method using arithmetic averages. All computations were performed using the microcomputer programme. NTSYS ver 1.6.

Mangifera has been divided into two subgenera by Kostermans and Bompard (1993). The first subgenus is called Mangifera and contains four sections. The second subgenus is

Limus and contains two sections. Additionally, there are many species of uncertain relationship. Kostermans and Bompard (1993) place the species M. casturi, M. indica, M. laurina, M. quadrifida and M. torquenda in the subgenus Mangifera, section Mangifera. Within this section, M. quadrifida and M. torquenda were considered to be more closely related to each other than M. casturi, M. indica and M. laurina. The other species discussed here belong to the subgenus Limus with M. decandra in section Deciduae and M. foetida, M. odorata and M. pajang in section Perennis.

In the present study, 109 bands were resolved using the ten primers selected. This phenogram does not support the morphological classification of Kostermans and Bompard (1993). The two subgenera appear to be mixed in the cluster analysis with M.quadrifida and M.torquenda clustering closely, and M. foetida and M. pajang branching from the same point.

The two sections were separated and analysis was run on each data set. The results suggest that within the Mangifera section, M. casturi, M. quadrifida and M. torquenda are closely related, but M. indica and M. laurina are quite distant from the others. Within the subgenus Limus, M. foetida and M. pajang are genetically close; M. decandra and M. odorata are more distant. These data support some of the groupings based on anatomical characters, but do not support relationships among others based on the morphological classification of Bompard and Kostermans (1992). This finding, therefore, emphasizes the usefulness of molecular taxonomic analysis in genome classification and genetic relatedness.

In the second study, RAPD markers were used to identify cultivars of Mangifera indica L. and validate genetic relationships among them. Within the species M. indica L. (the common mango), genetic relationships are difficult to determine because usually only the maternal parent is known. Controlled pollinations are difficult to make because of the small flower size. The long juvenility period of mango would make RAPD markers an extremely useful tool for identification of cultivars. This would also aid in the management of germplasm collections of mango, where identical cultivars often have different names.

Twenty-five clones of mango were selected for this study. Eighty primers were initially screened using 5 genotypes to determine the suitability of each primer for the study. Of the 80, 33 did not amplify, 19 amplified but were monomorphic and 28 amplified and were polymorphic. From the 28 primers that were polymorphic, 11 were chosen for screening of the experimental population.

The Percentage of Unique Pattern among 25 Mango Accessions for each Pairwise Combination of Primers

Primer	*OPA 15*	*OPA 18*	*OPA 19*	*OPA 20*	*OPF 6*	*OPF 12*	*OPF 13*	*OPF 15*	*OPF 16*	*OPF 20*
OPA 15	0.04									
OPA 18	1.00	0.92								
OPA 19	0.76	1.00	0.48							
OPA 20	0.60	1.00	0.92	0.28						
OPF 6	0.52	1.00	0.48	0.52	0.12					
OPF 12	0.64	0.92	1.00	0.92	0.76	0.08				
OPF 13	0.52	1.00	0.92	0.92	0.72	0.68	0.52			
OPF 15	0.36	1.00	0.80	0.56	0.48	1.00	0.80	0.16		
OPF 16	0.04	0.92	0.48	0.40	0.28	0.40	0.52	0.02	0.00	
OPF 20	0.68	1.00	0.84	0.84	0.92	0.92	1.00	0.76	0.64	0.48

It was found that primer OPA 18 was the most useful primer for identifying individuals, differentiating all but 2 of the 25 genotypes. Five other primers were also found to be effective. Thus with six primer sets, most of the common mango cultivars grown in the USA could be distinguished and characterized. In the case of genetic relationships, phenogram constructed based on RAPD data generally supported the known and assumed pedigree with two exceptions.

It was shown that the Indian cultivars 'Mulgoba', 'Langra', 'Vanraj' and 'Madras' differ from each other in many traits and from other Indian cultivars examined. Thus, in the day-to-day management of the germplasm collection, RAPDs allow identification of redundancy and provide an additional cultivar verification method.

Conclusions

The usefulness, reliability and potential of molecular markers in varietal identification, assessment of genetic diversity and establishment of genetic relatedness have been well-documented. These methods will definitely be the future techniques for taxonomic identities. With the development of RAPD markers, time and costs involved in such analysis have been considerably reduced.

The features of the technique make it an ideal candidate for automation. It is, therefore, not hard to envision the automation, integration and standardization of all steps in DNA fingerprinting methodology for the routine fingerprinting analysis of DNA samples and for genome mapping.

5

Genetic Principles

With many crop species, the variants of single genes (alleles) existing in subdivided populations are targeted to be sampled and maintained in genetic conservation. With clonal crops, the genetic integrity of specific genotypes, representing certain combination of genes which produce the desirable phenotypes, are often considered to be the basic units of genetic diversity to be collected and maintained. The following presentation attempts to reveal the basic genetic principles and terms related to these two approaches for sampling and maintaining genetic diversity.

Different Types of Reproductions

Tropical fruit crops can be reproduced sexually or asexually. Asexual reproduction is achieved by vegetative propagation, which with some species can also be made through in vitro techniques. The advantage of asexual reproduction is that the selected genotypes can preserve their genetic integrity; vegetatively produced progeny maintain also parental performance and uniformity. Sexual reproduction, production of seeds, can be based on self or cross pollination.

The level of self-pollination can range from zero to complete, although most crop species fall into two groups: those that are nearly always self-pollinated and those that are always frequently cross-pollinated. The majority of tropical fruit species are basically cross-pollinators, although variations between species and between genotypes within species occur. Near-obligate or obligate cross pollination may be imposed by separation of the sexes through either monoecy or dioecy. In monoecy, the pollen and ovaries are produced by different flowers on the same plant. With species showing dioecy, the male and female flowers are on different individuals. Obligate cross-pollination may also be imposed by systems of self-incompatibility where all flowers of a plant are perfect, but self-pollination will not occur due to inhibition of pollen produced by maternal plant. Cross-pollinated species have usually high level of heterozygosity, so single plant progenies are not true breeding *i.e.* genotypes differing from the parental genotype are produced through sexual propagation.

The imposition of self-pollination with naturally cross-pollinating species will lead to inbreeding depression, often also to failure to produce seeds and to expression of abnormalities. Naturally, self-pollinating species or genotypes have high level of homozygosity and consequently their progenies are often highly true breeding. Real self-pollinators do not show inbreeding depression.

The different types of reproductions are the basic element in developing the principles to maintain genetic diversity as alleles in divergent populations or as range of selected genotypes.

Sampling of Allelic Variation

Distribution of allelic variation in a population and between populations.

The number of distinct alleles per locus in a population, allelic richness, is used as measure of genetic diversity and the integrity attempted to be maintained. Since each gene contains may be from hundreds to thousands nucleotides, each a potential target of different types of mutations, the amount of allelic states at a single gene is virtually infinite; and the number of different allelic combinations of several loci are, if possible, even greater. Although the actual numbers of variants found or observed in a population are small in relation to the potential allelic variation, they are nevertheless very large.

The model of selectively neutral alleles (Kimura and Crow 1964) is most often used to estimate the expected number of alleles in a population. The application of this model is argued basically for two reasons. First, an extensive theory exists on the population/genetic structure of neutral variation and sampling of such variation. And secondly, the model yields the expected number of alleles based on single indexing parameter determining the distribution of allelic frequencies. This indexing parameter (q) is defined as: q = 4 Ne, where Ne is the effective population size in equilibrium and the mutation rate.

The important relationship between number of alleles in the sample and sample size is shown with the neutral allele model to be logarithmic, *i.e.* expected number of alleles increase in the logarithm of the sample size (Ewens 1972). Arguing that the more common alleles in the population are those of high priority, the allele frequency higher than 0.05 is often used as critical limit. With neutral allele model, more conservative values than 0.05 drastically increases sample size but have only minor gains in number of alleles caught into a sample (Marshall and Brown 1975).

To link the genetic divergence (allelic differences) between populations to evolution through natural or artificial selection, the most commonly used approach recognize four categories of alleles based on terms widespread vs. local and common vs. rare: (i) common widely distributed; (ii) rare widely distributed; (iii) common locally distributed; and (iv) rare locally distributed (Marshall and Brown 1975). Derived from the arguments that the first class (i) is easy to sample, the second (ii) depends only on the total number of plants taken, not the allelic distribution between populations and the locally rare alleles (iii) are those which can be far more effectively produced with artificial mutations than with searching for them at random in natural populations. It can be concluded that critical are those alleles which are locally common (iv) since they represent the variation which makes populations adapted to their environment or agricultural practices (Marshall and Brown 1975; Brown and Marshall 1995).

Sampling of Genotype Variation

In sampling of genotype variation, different genotypes or genets are targeted to obtain a range of phenotypes to be collected and maintained. Genet-approach is applicable for those species which can produce and maintain fixed genotypes with their mating system (strict self-pollinators and apomictic species) or are capable to clonal propagation. The number of different genotypes, genet richness, is the measure of genetic diversity applied to give the range of phenotypes, taking no account for the nature and extent of genetic differences producing the phenotypes (Brown and Briggs 1991).

The exact theoretical base for identification and sampling of genet richness is less developed than that for neutral allelic variation. Evidently, the only theoretical model for sampling of genetic richness is uptil now that of Brown and Briggs (1991) developed for the cases to conserve rare non-cultivated plants.

In practice, selective sampling, attempting to identify divergent phenotypes, is the method usually applied for clonal species including fruit trees. As it is impossible to conserve all possible alleles, it is also an impossible task to conserve all possible genotypes. Consequently, the users of selective sampling are forced to focus on germplasm which is known to be particularly important for further evolution of the species.

The sampling of representative sets of genotypes is the crucial task in conservation of genet richness of clonal species-genetic aspects related to maintenance of clones being more straightforward than in seed propagation.

Maintenance of Allelic Richness

Effective Population Size

The infinite neutral allele model is based on genetic theory developed for very large random mating populations. However, real plant populations are normally small in genetic sense (Spencer et al. 1991). With populations of crop plants different life history features, cultivation practices (selection of the seed material from harvested yield) and different environmental factors (*e.g.* soil fertility, rainfall, or temperature) can drastically influence to the number of breeding individuals in a population, *i.e.* the effective population size.

During maintenance, the effective population size has to be in accordance with the true population size, otherwise the difference between size of true population and effective population is a direct measure of genetic shift produced in a regeneration.

Allelic Divergence of Subdivided Populations

Genetic differences (allele frequencies) between populations, particularly those of long-lived perennials, are usually result of a long term evolution: alleles which are common in a particular population can be very rare, even be totally absent, in another population. As in sampling of allelic diversity, in maintenance of populations, alleles which are locally common' are critical since they represent the variation which makes populations adapted to their environment or agricultural practices.

In field genebanks, accessions collected from different environments are located for maintenance in one environment. What can easily happen is that a particular allele in a population or a certain population can have clear selection advance in the environment where genebank has been established. In genetic terms, a genetic shift in allelic frequencies within a population and between populations points towards the fittest allele and population can easily be produced.

Therefore, to maintain the genetic integrity of subdivided populations, the effective isolation distances have to be carefully studied before seeds to represent particular populations are produced. Populations which are poorly adapted to field genebank environment are the most critical and easily affected through outbreeding from other populations and they are also prone to different environmental bottlenecks and drift.

Maintenance of Clonally Propagated Genotypes

Clonal propagation fixes the genetic integrity of a genotype, therefore, compared to reproduction through seeds, the maintenance of clones is, in any sense, a more straightforward task. However, clonal genotype can be changed through spontaneous mutations and, if in vitro conservation methods are used, through somaclonal variation. Careful characterization, supplemented when possible with the molecular markers is, therefore, needed to monitor possible mutations during maintenance.

Genetic Aspects and Field Genebank

Genetic Bottlenecks and Random Genetic Drift

The single event producing sharp and drastic reduction in the population size is termed bottleneck. The initiation of a new populations from these few or even single selected or survived plants is called founder effect, since the structure of the new population is totally determined on the basis of the group of its founders. When a population remains small for several generations, cumulative sampling effects can result in changes in gene frequencies by chance alone, these chances can by no means be predetermined and the process is known as random genetic drift.

In maintenance of allelic variation, bottleneck and genetic drift have naturally strong effects on the genetic composition of a population both in qualitative (number of alleles) and quantitative (genetic variance) sense, but the immediate effects on the allelic variation are far more significant than those on the genetic variance per se. That is because rare alleles have only a minor influence on the variance but they are the most probable ones to be lost in a bottleneck. Though bottleneck is a single event, drift is not, the random chances of allele frequencies due to the sampling error in drift can be in fact, in a case in which population size is small through many generations, a series of random bottlenecks. Consequently the drift can have serious influence on allele frequencies as well as on genetic variance (Frankel and Soule 1981; Hartl and Clark 1989).

If only sampling of allelic variation in a population is considered, small populations do not basically differ from the large ones, the relationship between sample size and expected number of allele is shown to obey the same logarithmic function despite the effective population size (Brown and Briggs 1991). But populations after bottleneck, and especially those gone through serious drift, are naturally eroded in allelic variation. Thus, an appropriate minimum sample size to cover the most of alleles of small population can be in many cases much smaller than in the cases of large populations (Brown and Briggs 1991).

The methods of genetics are based on measured data taken from genotypes or populations. The data on genetically variable traits, genetic markers are, therefore, the basic element to combine genetics to genebank management. During different stages of conservation process, from acquisition to characterization and evaluation, different kinds of data related directly or indirectly to genetic differences between genotypes or populations are produced. This information can be exploited in many ways to strengthen the long-term conservation, to rationalize the maintenance of the collections and to improve the accessibility to the conserved germplasm to support its use in breeding programmes.

Different Kinds of Genetic Data

General Terms

Only traits expressing polymorphism, observable variation, have potential applications as genetic markers - increase in polymorphism ensures the increase in number of distinguishable genotypes. In order to make relevant comparisons between accessions, the genetic marker used should be homologous, *i.e.*, alleles of the same locus. If similar phenotypic features are non-homologous, they are termed analogous. Very often polymorphic traits are expressions of many different loci and distinguishing homologous from analogous markers is much more complicated by difficulty of recognizing serial homology or paralogy (Bretting and Widrlechner 1995).

With some markers, dominant homozygous and heterozygous genotypes can be distinguished and consequently these markers are said to be codominant to make them different from dominant markers. Mendelian gene is a term used to refer to gene expressing a morphological trait obeying rules of Mendelian inheritance, but it has to be distinguished from Mendelizing unit produced with new molecular genetics methods which is also discrete unit transmitted from parents to offspring and obeying rules of single gene inheritance, but which do not necessarily have anything to do with protein coding region. Observed phenotypes are usually products of both genetic and environmental factors; the relationship VG / VP is termed heritability (broad sense heritability in this case; VG is genetic variance and WP phenptypic variance). Highly heritable marker traits make the genetic analysis very much simpler than with traits expressing strong environmental effect and genotype-environment interaction (GE-interactions), term for the variable response of genotypes to changes of environments. Data with which the genotypes can be categorized to clearly inheritable similarity or dissimilarity groups is termed discrete data as distinct from continuous, normal-distributed data, typical for quantitative phenotypic traits (Falconer 1989; Weir 1996).

Morphological Traits

Highly heritable morphological traits, termed usually qualitative traits, are the commonly used genetic markers. Since the assay of qualitative traits do not need any sophisticated equipment or complex experiments, they generally are simple, rapid and inexpensive to score. They are normally dominant markers, a trait that prevents the use of morphological data in genetic studies in which allelic frequencies are needed. In general, the use of morphological traits as potential genetic markers requires the determination of the genetic control of these traits.

Quantitative traits, such as yield performance and quality characters, are the ones of major importance in breeding. Consequently, these traits are usually focused during the evaluation of accessions. Quantitative traits are usually normal-distributed and express strong environmental effects, often also genotype-environment interaction.

Allozymes and Seed Storage Proteins

The applicability of allozymes as genetic markers is related to their frequent polymorphism, codominance and Mendelian inheritance. Compared to methods for DNA-markers, protein analyses (electrophoresis) are also relatively simple and inexpensive. The major problems of storage protein or allozyme markers are that they are not randomly dispersed throughout the genome but just represent variation in certain specific loci encoding the given polypeptides. Mutations occurred in the loci are also detectable only as far as they affect the size or charge of encoded polypeptide and so produce changes in migration rate in electrophoresis (Simpson and Withers 1986; Gepts 1990).

DNA-markers

Compared to phenotypic or allozyme variation, DNA polymorphism has many obvious advantages. Heritability is generally at its highest since straight DNA variation is used. DNA-markers are also dispersed usually throughout genome's coding and non-coding regions. Furthermore, plant cells include three different genomes: nuclear (nDNA), mitochondrial (mtDNA) and cloroplastic (cpDNA), evolving with variable mode and speed and so allowing genetic analyses mostly unique for DNA markers. The three different genomes of plant which can be analyzed separately with these techniques, together with the other superior characteristics of DNA markers, have drawn out totally new prospects for different applications in genetics including those of genetic conservation. However, DNA analyses require sophisticated laboratory facilities. They are also relatively time consuming, laborious and expensive.

Some General Considerations for Analytical Methods for Genetic Marker Data

Data based on highly heritable morphological traits, allozymes (and other protein polymorphism in electrophoresis) and DNA-polymorphism are of major importance in studies of genetic similarities and differences in characterization and maintenance. Codominant markers (allozymes) RFLPs, SSRs), revealing allelic frequencies, are the most effective for studies of genetic diversity or genetic distance, although other type of markers can be used within certain limits. Quantitative traits are characterized with strong environmental effects on their expression and usually also with GE-interactions. Environmental effects decrease the heritability in general and GE-interactions can produce drastic differences between heritability estimates based on data taken from different environmental conditions. Therefore, the determination of genetic compositions of these traits requires experiments carried over time and space and are laborious and expensive. For these reasons, quantitative traits can be regarded to be quite unsuited for formal analyses of genetic diversity, genetic distance or genotype identification, and their use should be restricted to evaluation of germplasm.

Applications of Genetics to Field Genebank Management

Acquisition of Germplasm

Genetic marker data are instrumental in assessments of how completely a germplasm in collection represents a particular genepool targeted to conserve and consequently in assembling a collection with maximum diversity. Sampling and acquisition strategies used in establishment of field genebanks will be introduced in detail in other presentations.

Identification of Duplicates

The existence of unknown number of duplicates, *i.e.*, the presence of the same genotype registered as different accessions within a collection have been realized, to be a major problem in field genebanks. Such duplications increase unnecessarily the size of the collection making the conservation more laborious and expensive. Usually, attempts to identify duplicates have used only clearly inherited morphological traits. The confirmation of clear inheritance of the traits as well as the use of standardized list of descriptors are essential to make the results reliable and comparable between collections. With some species, duplicates have been identified by using step by step process applying different kinds of genetic markers (qualitative traits, quantitative traits, allozymes, DNA markers).

With long-lived woody perennials, such as fruit trees, biochemical techniques (DNA markers in particular) have clear advantages, since the establishment of morphological data is much more time consuming and expensive than with annual species.

Structure of Field Genebank

Establishment of a structure for field genebank is necessary to rationalize the acquisition, maintenance, evaluation and use of collected germplasm. Patterns are also the tool to asses collecting gaps and redundancies, to identify needs for resembling of germplasm and in establishment of conservation strategy (*i.e.* stratified use of different conservation methods including in situ conservation) for genetic diversity in genepool of a species.

The structure of a collection can be established based on different kinds of data related directly or indirectly to genetic similarities or differences between conserved genetic material. Data on highly heritable morphological traits, allozymes (and other protein polymorphism in electrophoresis) and DNA-polymorphism are the most suited to revel true genetic patterns (measures of diversity, population subdivision, genetic distance). Comprehensive and complete genetic data are seldom available or not reasonable to produce for whole collections. However, passport information can be used to establish ecogeographical patterns. In most cases, genetic differences among accessions are firmly linked to environmental pattern. Similarly, taxonomy and evolutionary/breeding history (*e.g.* wild forms, landraces, breeding lines, cultivars) can be used to produce genetically meaningful groups of accessions to improve the management of germplasm.

The concept of core collection was introduced to rationalize the evaluation of a collection and to improve the use of the germplasm. Core collection is a limited set of accessions derived from the existing germplasm chosen to represent the genetic spectrum in the whole collections (Brown 1989). For many species, core collections have been established using different kind of data including passport, characterization and evaluation data and different combinations of data (Hodgkin et al. 1995). Sets of accessions expressing certain desirable phenotypic traits (*e.g.*, 'disease resistance set', 'high-quality set', etc.) can also be formed to rationalize the use of collection. These subsets should not to be confused with the core collections, which are established to represent the whole genetic diversity existing in collection (Brown 1995).

Evaluation of Accessions

Evaluation of accessions is often the most time consuming and expensive task in conservation process. Cores, subsets and other structural components of the collection should primarily be established for exploitation of genetic patterns existing in collection for allocation of efforts in evaluation. Using these subdivisions of the collection, more systematic investigations can be focused on the parts of collections expressing certain desirable phenotypic traits. Quantitative traits which usually are in major focus are very often characterized with significant genotype-environment interaction, *i.e.,* different response of tested genotypes to the changes in environments.

Due to GE-interactions, it is impossible to predict the phenotypes in another environmental conditions on the basis of one location trials. Therefore, the analyses significance of GE-interactions (Falconer 1989) on the expression of a trait should be used in evaluation to improve the applicability of the results in the use of the germplasm.

Linkages to Use

Of the different forms of ex situ conservation, field genebanks are probably the most costly to establish and to maintain the requirements, especially of labour and supplies. Due to economic as well as practical problems in management of field genebanks and simultaneous urgent needs for crop improvement, it has become evident that firm linkages between conservation and use are needed for strengthening the sustainable long-term management of field genebanks. Genetic aspects discussed above are, therefore, not only a part of field genebank management but also essential in improving the access of conserved germplasm to users.

6

Propagation Techniques for Tropical Fruits

The propagation of plants is of great importance in perennial/vegetatively propagated crops. When new kind of plants have to be conserved or propagated or exploited, we need to develop knowledge and technique to propagate them. Invention of propagation structures and discovery of root inducing chemicals, in addition to grafting and budding methods, have revolutionized the propagation and nursery procedures. An appropriate propagation technology can be selected for each kind of plant based on its growth, physiology, flowering and phenology.

The propagation method can be primarily designed based on the mode of natural perpetuation like species with polyembryony can be propagated through seeds. Further, the general technique standardized for the species can be applied also to other species within the same genus with suitable modifications. This could be well explained within the family Anacardiaceae where mango and cashew both can be successfully multiplied by epicotyl grafting and soft wood grafting. This paper gives details on the propagation methods followed in some of the major and minor tropical fruits of the Asia-Pacific region, such as mango, citrus, rambutan, jackfruit, litchi, durian and mangosteen.

Propagation Techniques

Mango: Mango is a cross pollinated and highly haterozygous fruit crop. Therefore, propagation by seeds (called stones in mango) lead to variability in the progeny and is a limitation for commercial orcharding, though desirable for breeding new varieties. Hence, vegetative methods of propagation are adopted for getting true-to-type plants. Seed propagation is still the chief method of multiplication of rootstocks.

Propagation of Rootstocks

Seed Propagation: Rootstock plants are raised from stones. Most of the commercial varieties in India are monoembyronic but some varieties from South India are polyembryonic which give true-to-type seedlings from nucellar embryos (Sachar and Chopra 1957). Stones collected indiscriminately give poor germination but provide a lot of variation. Bolder stones give seedlings with more vigour irrespective of whether these were collected from grafted or seedling trees. The viability of mango stones is lost very quickly. Freshly collected stones from canning units give high germination but when these seeds are stocked or kept in sun for a large period, they lose germinability.

The germination of mango stones is 80% when sown within a month of extraction but none thereafter. The viability of stones could be extended by storing them in charcoal powder (Teaotia and Singh 1971) and polythene bag (Singh and Kaisuwan 1971). Patil et al. (1986) reported 40% germination after 90 days of storage in polythene bags with charcoal. Although refrigerated storage had adverse effect on germination, seeds were successfully preserved for 12 months without any decline in viability and vigour after treatment with 1% hydroxyquinoline sulphate and storage in polythene bags of 100 guage at 15° C (Doijode 1989).

The selected stones are spread on raised nursery beds and covered with leaf mould or FYM for better germination. Germination starts within 12 - 15 days and continues up to 3 weeks. The germination is adversely affected if the stones are sown too deep or in pots or polybags. Seedlings are transplanted to pots or polybags as soon as they turn green and these are nursed and prepared for grafting. When polyembryonic seeds are sown, nucellar seedlings are separated within a month and planted separately.

Vegetative Propagation: Propagation of rootstocks by vegetative methods is advocated to eliminate variation. These methods are not commonly used probably due to absence of standard rootstock type. (1) Cutting: The hardwood leafy cuttings are used for rooting. Cutting from younger trees rooted better than those from old trees (Mukherjee et al. 1966). Rooting as well as subsequent survival has been improved by using juvenile shoots, etiolation and ringing of shoots and application of IBA in lanolin paste on the ringed portion (Bid and Mukherjee 1972). A medium containing peat moss and sand (1:1) proved better and such cuttings have to be kept under mist for rooting.

Use of bottom heat to provide a temperature of 30 to 20°C (Reddy and Majumdar 1975) for Deshehari cuttings dipped in 5000 ppm IBA and planted in sphagnum moss and sand mixture resulted in 97 per cent rooting compared to only 15 per cent without bottom heat. Beneficial effect of bottom heat was observed by Rajan and Ram (1983), Sadhu and Bose (1986), and Reddy and Singh (1987). The synergism of phenols and flavonoids with IBA for root regeneration in cuttings has been demonstrated by Reddy and Majumdar (1975). Preplanting treatment with phenolic compounds such as hydroxybenzoic acid, coumaric acid and ferulic acid generally promoted auxin-induced rooting in cuttings (Sadhu et al. 1978).

Air Layering: Etiolation of shoots before layering resulted in more fibrous roots than without etiolation and application of IBA + NAA improved rooting (Bid and Mukherjee 1969). The success of air layers was markedly improved by use of growth regulators like 5000 to 20 000 ppm NAA and IBA in lanolin paste (Chhonkar and Singh 1972).

Stooling: Majumdar and Mukherjee (1961) described modified form of stool layering, wherein after about two years of growth, the mother plant is headed back at 10-12 cm above ground to induce emergence of several shoots below the cut end. Vigorous shoots are ringed and IBA (5000 ppm) in lanolin is applied on the ringed portion about a week before earthing up. Rooting takes place in 4 to 6 weeks and these are separated from the mother plant. The same mother plant can be utilized for about ten years to obtain genetically uniform rooted plants of a given clone. Clonal propagation of mango rootstocks by stooling has also been reported by Singh and Srivastava (1982). This is the easiest and fastest method for vegetative propagation to multiply mango rootstocks.

Propagation of Scion Plants

Mango plants in nurseries are raised by vegetative methods like budding and grafting.

Budding : Various budding methods, namely, patch forkert, shield, chip and T-budding are successful in mango. However, the success varies depending on geographical location and season. Since the buddings are likely to be damaged during transport, this method is not followed commercially.

Grafting : Inarching or approach grafting, veneer grafting, wedge or whip grafting, epicotyl grafting or stone grafting and softwood grafting are the most extensively used methods.

(1) *Inarching:* This is the oldest and still the chief technique employed in many mango growing regions. Although, it gives very high success, this method is costlier, cumbersome and requires larger bud wood and a long time to produce grafts.

(2) *Veneer Grafting:* In this method, mature scion of 3-4 months is grafted on 1 to 2 years old stock. Pre-curing the scion shoot by defoliation atleast 7 to 10 days prior to grafting is essential. However, in the coastal areas of Maharashtra, prior defoliation is not required (Gunjate et al. 1976). The period of veneer grafting differs from place to place owing to climatic variations. This technique is suitable for in situ grafting and for top working of seedling trees.

(3) *Wedge Grafting:* Mature procured scion shoot of 3 to 4 months old is wedge grafted on 1 to 2 year old stock. The success is high during the humid conditions and 90 to 100% success is obtained under greenhouse and mist chamber. This method is also used for in situ grafting and top working.

(4) *Epicotyl Grafting or Stone Grafting:* In this method, mango seedling of less than two weeks is wedge grafted with

mature scion. For higher success and better growth, use of two stocks per scion is recommended. This is the commercial propagation method in Konkan region of Maharashtra and does not require pretreatment of scion (Gunjate et al. 1982).

(5) *Softwood Grafting:* In this technique, grafting is done with mature, precured scion on the emerging soft copper red shoot as described by Amin (1974). This technique was successful in establishing mango orchards in situ in drier tracts where mortality of nursery raised grafts is very heavy (Amin 1978).

Citrus

The sexual as well as asexual (vegetative) methods of propagation are used to raise plants of different species in citrus. The seedlings of acid lime and mandarins are still planted in southern and northeastern region of India.

Seed Propagation : The number of seeds per fruit varies between different citrus species and cultivars. Seeds obtained from healthy, virus-free old trees which have a pedigree performance of producing vigorous, uniform seedlings should be used.

Polyembryony : Seeds of most citrus are polyembryonic and thus nucellar seedlings are used both for raising uniform rootstocks as well as for direct planting, especially in acid lime and mandarins (Singh and Singh 1955). This also helps to raise healthy plants as citrus viruses are not transmitted through seed.

Gopalakrishna and Kunte (1958) reported varietal differences in the degree of polyembryony in citrus. Motial (1963) observed that the number of embryos per seed was 4.7 in Jambhiri, 4.45 in Italian lemon and 4.25 in Galgal with a range of 1 to 11 in Citrus species. Galgal and Sadaphal had 90 and 95 per cent polyembryonic seeds respectively. Singh (1965) opined that constricted surface of seeds in Kama Khatta, Citrus jambhiri, C. aurantium and Poncirus trifoliata was associated with polyembryonic nature, except in P. trifoliata with polyembryony varying from 35.5 to 54.2 per cent. In citrus, deformed or uneven seeds were found to be usually polyembryonic (Singh 1971). Prasad and Ravishankar (1983) claimed 100 per cent polyembryony in Rusk Citrange, Trifoliate Orange and Swingle Citrumelo (93.3%), Troyer (86.6%), Carrizo (73.3%), but there was no relationship between the percentage of polyembryony and seed morphology.

Viability : Since Citrus seeds generally have no dormancy and if they are allowed to dry and the cotyledons separate, the seeds fail to germinate. Therefore, it is essential to sow the seeds immediately after extraction. The viability during storage varies depending upon species and storage conditions.

Quick evaluation of seed viability could be done by tetrazolium test soaking the seeds in 1 % triphenyl tetrazolium bromide solution which strains the living tissue (Singh and Chopra 1962).

Seed storage of Kagzi lime and Mosambi in charcoal powder at 50 per cent RH at room temperature was better than sealed polythene bags (Bajpai et al. 1963). Kumar and Majumdar (1973) reported that in C. aurantifolia, the seed germination was highest in freshly extracted seeds and declined as the storage period extended. Krishna and Shankar (1974) observed that in C. karna and Rusk Citrange, seeds viability was 100 per cent after 150 days storage in polythene bags with calcium chloride at 8°C, whereas in case of C. jambhiri and C. limonia, it was 84 and 80 per cent respectively.

Raising Seedlings : Citrus seeds are to be sown on raised and well manured beds at 2.5 cm apart in rows spaced at 15 cm. In commercial nurseries, sowing is done during mid August to February. Singh et al. (1970) observed that alkathene cover improved seedling growth of C. jambhiri, C. pseudolimon, C. limonia, C. megaloxycarpa and P. trifoliata over those in open. Many workers reported the usefulness of soaking the Citrus seeds in GA from 200 to 600 ppm to improve the rate of germination (Shant and Rao 1973; Mishra et al. 1982; Singh et al. 1989).

Vegetative Propagation

Cutting : Many species of Citrus can be successfully raised from cuttings and it is a very useful method of propagation especially when a species is desired to be colonally propagated on its own root system. However, there is wide variation in the rooting ability of different Citrus species. While limon and citron are easy to root, mandarin oranges, sour orange and grapefruit root with considerable difficulty (Sadhu 1986). Synthetic auxins (IBA, NAA and IAA) and phenolic compounds help in rooting of cuttings (Dhatt and Zora Singh 1993). However, age and physiological status of mother plant, type of wood, time of planting and media composition for planting of cuttings determine the extent of success (Gangwar and Singh 1965; Bajwa et al. 1977).

Air-layering : Air-layering is fairly common and is used to propagate pummelo, lime and sweet lime. Singh (1955) achieved appreciable improvement in the rooting of air-layers of sweet lime, grapefruit and sour lime by treating the cut surface with auxins (NAA + AAA both at 1 per cent). Shankar (1965) noted that certain rootstocks like C. karna, C. jambhiri, C. pennivesiculata, C. assamensis and C. megalocarpa can easily be raised from the air-layers even without treatment of growth regulators.

Budding : Budding is the most common and widely used method of vegetative propagation of Citrus. T-budding is probably the best method and is successful with many species of Citrus. Naik (1939) exhibited that bud take in Citrus spp. increased when wood was not removed from the bud. Stevenson (1947) reported that 'T'-budding gave 87% success in Citrus species. In Punjab, the bud take of Kinnow on C. jambhiri was 98% with better subsequent growth when round buds without wood were used than by using angular buds with or without wood (Sharma and Sharma 1986). The time of budding affects bud take and subsequent growth of-plants. Experiments conducted at different places in India have shown that February and September-October are the best periods for budding in most Citrus species (Dhatt and Zora Singh 1993).

Now polythene strips are commonly used as wrapping material instead of jute twine. Srivastava and Arya (1969) observed, while budding mosambi on C. karna, that 200 guage polythene strip was the best wrapping material. Delayed lopping of the rootstock till the bud has grown about 4-5 cm is usually practised by nurserymen.

Rambutan

The rambutan is one of the best known fruits of South-East Asia and is especially important in Thailand, Malaysia and Indonesia, but extends to the east as far as the Philippines.

Seed Propagation: Propagation by seed has been the traditional practice and large number of trees now in production throughout South-East Asia originated from seed. This is not a desirable method since plants from seeds vary and many seedlings produce poor fruits. In addition, seedlings vary in sex. Male trees are practically useless, and female trees, unless well pollinated, do not bear well. Populations of seedlings of rambutan have yielded up to 67% males. Propagation from seed should only be used for the development of new varieties.

Viability: Seeds, when removed from the fruits, have a short life time and are recalcitrant (Chin 1975; Teng 1977). It is desirable to maintain them slightly moist and to plant them in 2 or 3 days. If the seeds are sown after 3 to 4 weeks, seeds usually do not germinate.

Raising Seedling

Seeds can be sown in seed beds or in individual containers. The latter method is preferred as it eliminates the risk of transplanting and shows better establishment in the field.

Vegetative Propagation: Vegetative propagation is indispensable for production of trees of high quality. These trees tend to come into fruiting at a younger age than seedlings. To preserve good, hermaphroditic varieties, vegetative propagation is necessary (Valmayor et al. 1961).

Air-layering: This is frequently used by home gardeners and small scale producers, but success rates are generally variable.

Grafting: Methods used include various forms of approach grafting and cleft grafting, which is particularly used in the Philippines. Top working of seedling or staminate trees and grafting with improved cultivars has been described.

Budding: In most of the South-East Asian countries, seedling stocks are budded by the modified Forkert method. Various other budding methods have been used such as shield or T- budding and patch budding with limited success.

Jackfruit

Seed propagation is an old method of multiplication which is planted in situ or raised in pots. Now, vegetative propagation is the commercial method of jackfruit multiplication.

Seed Propagation: This method is mainly used for the production of rootstocks. Soaking of seeds in NAA 25 ppm for 24 hours resulted in highest germination percentage as well as good seedling growth (Bose 1986). Shanmugavelu (1971) recorded 100 per cent germination by treating the seeds with GA3 at various concentrations upto 500 ppm compared with 80% in the untreated control.

Litchi

Seed Propagation: This method is not desirable since the trees raised from seeds have long juvenile period, fail to produce true-to-type plants and often produce fruits of inferior quality. Propagation by seed is used mainly for rootstock propagation and breeding. The seeds of litchi have a very short viability, usually less than 5 days.

Vegetative Propagation

Cutting : Propagation by cuttings is possible, but the plants produced are less vigorous and this method is not generally recommended (Abutiate and Nakasone 1972; Kadman and Slor 1974; Paxton et al. 1978).

Air-layering : The main commercial method of propagation is by air-layering and rates of success may be as high as 95% under ideal conditions. Air-layered trees normally produce commercial crops after 3-6 years (Menzel 1991).

Grafting : Due to the limited and unpredictable activity of the cambium, which is related to the onset of leaf flushing, grafting is not always very successful. There is also possibility that some scion/stock combinations may be incompatible. However, approach grafting is generally successful. Side grafting has been reported to be successful up to 80% in Israel (Kadman and Slor 1974).

Durian

Seed Propagation : Seeds germinate within three days after sowing and there is no dormancy. Percentage of germination reduces drastically when seed moisture level depletes beyond 22%. Seedlings take longer time (8-15 years) for fruiting.

Vegetative Propagation : Budding, grafting, air-layering or even cuttings are the vegetative propagation methods followed. However, whip grafting or patch budding are the most commercially used propagation methods. Better results of grafting/budding are obtained if rootstock of same species is used. Grafted or budded plants come to fruiting in less than 4-5 years.

Mangosteen

Seed Propagation : The best known practice for the propagation of mangosteen is by seeds and when fresh seeds are sown, they germinate in 10 to 15 days. Since the seeds are of asexual origin, they produce trees identical to the mother. If the seed is dried or kept outside the fruit for several days before planting, germination is drastically reduced. However, the seeds can be maintained for 3 to 4 weeks within the fruit. The best way to carry seeds is by transporting the entire fruit.

The size of the seeds is highly variable. The difference in seed weight brings variation not only in germination but also in later growth and survival. The ability to germinate and grow successfully is related to the amount of food stored in the seed. Therefore, it is recommended to establish new plantings only from the larger seeds, those that weigh 1 g or more.

Vegetative Propagation : Many attempts have been made to develop stronger, more rapidly growing mangoseen trees with a short juvenile phase through cutting, air-layering, Forkert budding, approach grafting and cleft grafting but the results were not promising (Campbell 1966). All the propagation methods tried did not show any real advantage over the seed propagation method.

7

Tissue Culture Strategies of Tropical Fruit Crops

A large number of tropical and subtropical fruit crop species have highly recalcitrant seeds often exhibiting viability of just a few weeks. Several vegetatively propagated tropical fruits including woody perennials need to be conserved with their genetic fidelity intact. Maintenance and preservation of germplasm of fruits in field genebanks is difficult, expensive and labour-intensive. Besides, the conection may also be vulnerable to natural perturbations.

Although conservation of seeds of economically important crops has been practised since times immemorial, it suffers from severe limitations like low seed viability and heterozygosity. Therefore, a need was felt to develop alternative methods, viz., in vitro preservation (Kartha 1981). New techniques help to conserve germplasm disease-free, but also involve lower labour costs and requirement for technical personnel besides limiting disease-transfer (Withers and Engelmann 1990).

Role of Plant Tissue Culture

The term plant tissue culture denotes genetically all cell, tissue and organ cultures, and employs the basic dogma totipotency. The technique essentially involves separation of a cell/tissue/organ from the donor plant under aseptic conditions and growing it on a synthetic medium in a suitable container under controlled environment. Among other applications, tissue culture has been successfully dovetailed to conservation and exchange of germplasm in horticulture.

In the past decades, in vitro techniques have been extensively developed for over 1000 different species. Tissue culture systems allow for axenic propagation of plant material with high multiplication rates. Virus-free plants can be obtained through meristem culture in combination with thermotherapy. The miniaturization of explants reduces space requirements and consequently labour costs for maintenance of germplasm collections.

It is now mandatory to exchange coconut germplasm across countries using embryo cultures because of the obvious advantages. However, it may be noted that mere transfer to in vitro culture does not automatically confer the disease-free status. Specific pathogenicity tests need to be carried for certification of the in vitro plant material as being disease-free.

Methods of in Vitro Conservation

Prolonged maintenance of in vitro material also provides an effective system for establishing both active and base germplasm collections particularly for the conservation and maintenance of those plant species that cannot be stored as true seed and are amenable to micropropagation technology (Fay 1994). The properties required of a successful storage system are the ability to:

- Reduce the growth rate of vitroplants, resulting in extension of intervening subculture frequencies.
- Retention of vitroplant viability during storage at maximum simultaneously maintaining genetic stability at minimum.
- Maintain the full developmental and functional potential of stored material when it is returned to physiological temperature.
- Make significant savings in labour input, materials and commitment to specialized growing facilities.

There are three types of in vitro storage for conservation:

1. Normal in vitro culture propagation.
2. Reduction in growth rate.
3. Suspending the growth.

Therefore, in vitro methods are different from either the in situ conservation in natural habitat or ex situ conservation in field conditions. Tissue cultures may be conserved at either relatively low temperature of 20 to 15°C (slow growth) or cryopreserved in liquid nitrogen at -196°C (no growth). Accordingly, In Vitro Active Gene Banks (IVAG) and In Vitro Base Gene Banks (IVBG) may be established, respectively.

Short to Medium term Storage for IVAG

The basic principle underlying this form of storage is the induction of reduced vegetative growth of the stored material by inducing osmotic stress, limiting the availability of carbohydrate to sub-optimal levels, low temperature/dark maintenance of cultures or incorporating growth retardants in the culture medium. These methods may be used singly or in combination.

Long-term Storage or Cryopreservation for IVBG

In this method, cultures are suspended at very low temperatures (-196°C) in liquid nitrogen to facilitate the arrest of mitotic and metabolic activities. The procedure guarantees long-term preservation of germplasm in genetically unaltered state. The chief drawback, however, is that the viability of very few biological materials can remain unaffected when they are frozen at very low temperatures, Cryoprotectants like dimethyl sulphoxide (DMSO) are used (Tables 1-2). Cryogenic strategies for survival of cultured plant cells are detailed by Kobayashi et al. (1991) and Sakai (1993).

Effect of Cryoprotectant on the Survival of Nuclear Cells of Navel Orange Cooled to -196°C by Slow Freezing Method

Cryoprotectant	*Survival (%)*
Control (No Additives)	30 ± 1.8
5% DMSO + 0.7M Sorbitol	30 ± 1.6
5% DMSO + 0.7M Glucose	30 ± 1.2
5% DMSO + 0.7M Sucrose	51 ± 1.2
5% DMSO + 0.7M Glycerol	3 ± 0.8

Nucellar cells were frozen slowly to -40° C at 0.5°C/min.

prior to immersion into LN. Values are means of triplications ± SE (Kobayashi et al. 1990).

Choice of Plant Culture System

The basic requirement for in vitro conservation is the availability of highly reproducible protocols for either micropropagation or regeneration as the case may be.

There has been a strong, though not universal, prejudice in the past in genetic conservation programmes favouring the use of shoot cultures in particular. This is presumably because of the greater stability of such cultures compared to the adventitiously propagated systems. All culture systems from protoplasts through cell suspensions, callus, somatic/zygotic embryo to shoots/nodes and entire plantlets have valid and appropriate roles.

Effect of Sucrose Concentration in Combination with 5% DMSO on the Survival Rate of Nucellar Cells Cooled to -196° C by Slow Freezing Method

Cryoprotectant	*Survival (%)*
5% DMSO + 0.5 Sucrose	21 ± 1.4
5% DMSO + 0.7M Sucrose	44 ± 1.8
5% DMSO + 1.0M Sucrose	63 ± 1.6
5% DMSO + 1.2M Sucrose	73.± 1.4
5% DMSO + 1.5M Sucrose	29 ± 3.0

Values are means of duplicates ± SE (Kobayashi et al. 1990).

Minimal Growth Storage:

Objective (Grout 1995)

- To devise an effective and reproducible protocol for minimizing the growth and development of plant materials.
- Avoid induction of abnormal development during storage.
- Preserve the viability of the stored population of in vitro cultures.
- Ensure that the protocol allows for periodic growth under standard conditions between periods of prolonged storage.

Growth at Reduced Temperature:

- Standard culture medium in suitable containers, vessels must be robust, semi- solid medium.
- Controlled temperature and reduced lighting, light intensity can be reduced by 60% from the standard growth level.

Procedures:

1. Prepare culture vessels with appropriate standard medium.
2. Establish fresh explants from active in vitro material growing at optimal temperature.
3. Characterize the cultures.
4. Incubate the cultures to be stored at their optimal temperature.
5. Transfer the cultures to the storage temperature (4-10°C).
6. Establish at maintenance level light regime with 16-h photoperiod (at 500-1000 lux being widely accepted).
7. Inspect cultures weekly, until satisfactory conditions have been confirmed.
8. At a preselected point, the cultures should be taken from the storage conditions, transferred to the growth medium and returned to optimal growth conditions.

The Addition of Growth Regulators: Materials: Abscisic acid, CCC (Cyclocel), etc.

The Addition of Osmotic Regulators

Material: Mannitol or sorbitol.

Combined Treatments : Combination of the treatments outlined above may be of benefit.

Cryopreservation of excised meristems, shoot tips and somatic embryos (Grout 1995)

This area has not been extensively researched. There are trends and protocols with promise that have been successful for a number of species. The preferred option for extended storage of meristems is Cryopreservation and yet in this respect, the excised meristem poses particular problems due to its mass and multicellular diversity.

Objectives:

- Preservation of optimum viability of the in vitro materials when recovered from liquid nitrogen.
- To ensure that plant regeneration from viable explants following Cryopreservation is by outgrowth of original frozen thawed meristems.
- To provide the maximum genetic stability for stored explants meristem/shoot- tips Cryopreservation by slow cooling.

Whilst there is no single procedure in this category that has universal application, there is a general approach that appears to be successful. This is:

- Pregrowth of donor plants or in vitro cultures at low temperature to bring about cellular changes associated with acclimation to low temperature.
- Pre-treatment of the excised meristem.
- Treatment with cryoprotectant.
- Cooling rate of 1°C per minute or less to intermediate temperature (typically - 40°C) followed by plunging into liquid nitrogen.
- Thawing by immersion in water at 40°C.

Procedures:

1. If the donor source of the meristem/shoot tip explants is likely to acclimate to cold, then a suitable hardening regime is to be devised When the donor material is an in vitro plantlet, a regime of temperature of day (22°C) and cool night (-1 to 4°C) may be applied for a period of 7-28 days.
2. Following aseptic excision of the meristem or shoot tip, a pre-treatment before cryoprotectant exposure and freezing is generally beneficial.
3. A suitable cryoprotectant treatment may be chosen.
4. Cooling is controlled from 0°C to an intermediate temperatures of -35 or -40°C, at a rate between 0.1 and 1.0°C per min, with rates between 0.5 to 1.0°C per minute being preferred.

5. During storage, the level of liquid nitrogen storage vessels be regularly monitored and LN replenished.
6. Vials are thawed by immersing them individually in sterile distilled water warmed to 40°C.
7. Once thawed, the tissue is transferred to the appropriate regrowth medium.
8. There is no viability test that can be employed to assess the success of Cryopreservation of meristem/shoot tips immediately after thawing.

By rapid cooling also meristem/shoot tip can be cryopreserved, somatic embryo, can be cryopreserved by dry freezing/rapid cooling/desiccation and with/without alginate encapsulation (Grout 1995).

In short, cryopreservation techniques are of two types:

(i) Protocols/procedures having a high technology input, such as use of a programmable freezer, or methanol cryobath, etc., which enables slow cooling at variable rates upto -30 or -40°C, followed by rapid plunging in LN.

(ii) Vitrification based procedures have, in comparison to that explained in (i), low technology input, with practical advantages. They are user friendly, operationally less complex and do not require processing of samples and slow cooling rates, thus avoiding use of sophisticated equipment like programmable freezer, methanol cryobath, etc.

In Vitro Germplasm Exchange

The importance of national and international germplasm exchange in vitro can hardly be over-emphasized considering the tremendous advantages it offers over conventional methods. Besides occupying far less cargo space, it also affords exchange of axenic plant material free of undesirable contaminants. This is of special significance in plant movement, quarantine and to breeders who can import a large number of clones of choice, multiply them and distribute them among user groups. In vitro exchange is, therefore, catching on fast. The principles are applicable to perennial fruits. It is important to use suitable, impact resistant and well sealed culture containers. The packaging should provide adequate thermal insulation and protection against rough handling. The culture medium should be of a formation that includes a higher than usual concentration of gelling agent. The most rapid available shipping method should be preferred.

Conclusions

A steady and progressive genetic erosion has resulted in a marked decrease in plant diversity. A wide genetic base helps to contain biotic and abiotic stresses that destroy entire crop stands. Developing countries, particularly those in the tropics, are still the repositories of vast biodiversity in their agro-ecosystem and act directly or indirectly as reservoirs and suppliers of germplasm.

Therefore, there is an urgent need to establish genestocks which will provide recruiting grounds for gene-hunters specially for crop varieties that run the risk of elimination by intensive agriculture. An adequate genetic resource conservation programme is to genetic engineering as a library is to knowledge. In this context, the tropical and subtropical fruit crop species stand to gain immensely from in vitro techniques of conservation and exchange of germplasm.

Although many genebanks exist worldwide today, only about 30 countries provide secure long-term storage because there is little provision for long-term sustainable management of genebanks. Therefore, an integrated approach is both prudent and practical. Success stories include tropical species such as cacao, coconut, Citrus spp., coffee, etc. Even so, in vitro conservation in tropical fruit crops has a long way to go and avenues for research in this area are wide open.

Pollen Storage in Tropical Fruits

Genetic conservation through pollen storage is desirable for a variety of horticultural plant species, since pollen is known to transmit important genetically heritable characters. Pollen is a product of genetic recombination and can provide a reliable source of nuclear genetic diversity at the haploid stage. Although genetic conservation through pollen storage does not accomplish the whole genome conservation, a plant breeder involved in genetic enhancement of a given horticultural crop could have access to a facility called Pollen Cryobank, from where he can draw pollen parents of his choice in the process of breeding a new cultivar.

Pollen genepools can provide sources of resistance for exchange with other breeders (Poulos 1994). Exotic nuclear genetic diversity can be easily received and exchanged through pollen, thereby eliminating the need to go through a long juvenile phase, common in most fruit trees to produce pollen for hybridization at a desired location. Thus, stored pollen can be used to improve breeding efficiency. Plant breeding strategies could derive the benefit of pollen selection when samples having increased longevity are indexed as viable and fertile for carrying out breeding operations throughout the year (Hanna and Towill 1995).

Fruit tree pollen is generally required to be stored for controlled crossings, either to achieve a desired breeding objective, or to overcome a constraint involved in commercial fruit production. For example, in Citrus rootstock breeding programme, interspecific or even intergeneric crosses are attempted to introgress genetic traits into the desirable cultivated plants from pollen parents known for providing resistance to biotic stresses. Citrus and its allied species flower asynchronously and successful crosses are seldom accomplished with fresh pollen due to non-availability of the seed parent when the pollen parent is in full bloom. Pollen storage has come to the rescue, where stored pollen indexed as viable can be used in crossing with the desired female clone so as to accomplish the breeding objective.

Successful transfer of trifoliate leaf character to the F1 zygotic hybrids of an intergeneric cross involving Poncirus trifoliata as the pollen parent and Citrus limonia as seed parent, using one year cryopreserved pollen (Rajasekharan et al. 1995) was accomplished. On the same lines, pollen of mango cultivars were successfully cryopreserved for two years (Rajasekharan and Ganeshan 1992). Pollen fertility was retained for five years as tested by intervarietal crosses.

Genepool conservation at the haploid stage can, therefore, be effectively accomplished through pollen which can provide a rich source of nuclear genetic diversity. A major emphasis on research needs include pollen storage (Arora and Rao 1995) for citrus, mango, rambutan, jackfruit, durian and litchi. Pollen cryopreservation research has been recently recognized by IPGRI (IPGRI 1995).

As one of the genepool components in an integrated PGR conservation programme, pollen can serve as an alternative or additional ex situ conservation method. Alexander and Ganeshan (1993) have extensively reviewed pollen storage research in fruit crops. Grout and Roberts (1995) have elaborately described the methodology involved in pollen cryopreservation. Hoekstra (1995) has assessed the merits and demerits of pollen storage for genetic resources conservation. The present lecture will give an insight on research progress made at IIHR with respect to citrus and mango pollen storage.

Citrus

Pollen Collection : The species handled for pollen collection include Citrus limon, Citrus aurantifolia, Citrus sinensis and Poncirus trifoliata.

Inventory : Petri dishes, butter paper, forceps, muslin cloth fixed to a 10 cm. cylindrical ring with the help of a firm rubber band, clean razor blade, etc.

Procedure : Pollen collecting are usually made on a bright sunny day between 8 - 10 AM, staminate flowers are harvested at peak anthesis and the dehiscing anthers are gently caressed over a muslin cloth sieve. Anthers are carefully removed from pistillate flowers and pollen is extracted over a muslin cloth sieve. Thus, pure pollen is extracted in clean petri plates or butter paper.

Precautions : The following points should be noted:

1. Quality of pollen collected depends on the correct stage of anthesis/anther dehiscence.
2. Pollen should be free of anther debris.
3. Do not collect pollen from infected or insect damaged flowers.

4. Decide upon bagging of inflorescence depending on insect activity.
5. Do not venture to collect pollen on a rainy day or if it had rained overnight.
6. Do not force out pollen from anthers.

Viability Testing : Pollen collections are subjected to viability indexing by germination in vitro by the hanging drop technique (Shivanna and Rangaswamy 1993). This process involves suspension of pollen grains in a drop of liquid medium consisting of a carbohydrate source, inorganic and organic compounds.

Inventory : Cavity slides, cover glasses (22 mm square), microslides, petroleum jelly, large petri plates, filter paper discs, incubator, double distilled water, needle, forceps, "Quick fix" (epoxy resin).

Medium : Prepared in deionised, double distilled water. 20% sucrose, 100 ppm H_3BO_3, 300 ppm $Ca(NO_3)24\ H_2O$, 200 ppm Mg $SO_4\ 7H_2O$, 100 ppm KNO_3, pH 7.3.

Procedure : Apply petroleum jelly to the edges of a new cover glass. Place this cover glass on the work table so that the edge on which petroleum jelly is applied faces the roof. Place a 25 to 50 l drop of nutrient medium on the centre of the cover glass assuming the shape of a small bubble. Add pollen grains with the help of a needle carefully filling the drop with optimum quantity of pollen. Invert a cavity slide over the cover glass with the cavity over the drop.

The petroleum jelly helps to adhere to the cover glass around the cavity, sealing the cavity from all sides. Reverse the cavity slide and place in a petri plate fixed with moistened filter paper disc cut to its exact size. Place the cavity slide over a microslide in such a way that the cover glass with the hanging drop over the cavity is well clear of its surrounding area, intact and hanging. Cover petri plate with the top lid (also fixed with moistened filter paper) and keep in an incubator at 25 ± 2°C, dark, for 18 to 20 hours.

After the prescribed duration of incubation, the hanging drop preparation is separated carefully from the cavity slide.

After removing the residual petroleum jelly from the slides of the cover glass, a drop of Alexander's differential stain (Alexander 1980) is added over the stained drop and mixed slowly. Place a clean microslide over the stained drop. The drop is flattened with the cover glass sticking to the underneath of the microslide. Remove excess stain and seal the edge of the cover glass microslide interface with a thin layer of quick acting adhesive. The slide with germinated pollen is ready for observations under the microscope for qualitative and quantitative viability estimates.

Precautions : Noteworthy points are:

1. Prepare medium fresh.
2. The medium on the cover glass should be just optimal for it to hang on the cover glass when inverted.
3. Add just enough stain to the drop containing germinated pollen. If excessive stain is added, the germinated pollen drifts away outside the cover glass when the drop is flattened, leading to errors in quantitative estimation.

Storage

Inventory : Butter paper, glass vials, silica gel, gelatin capsules, polyethylene aluminium laminated pouches, head sealer, domestic refrigerator, freezer, liquid nitrogen storage, cryoflask filled with liquid nitrogen.

Procedure : Pollen samples are packed in glass vials/ gelatin capsules/butter paper packets depending on the method of storage. Lemon pollen contained in glass vials are best stored at 5 or -18°C for short duration of 8 weeks (Ganeshan and Sulladmath 1983) and medium term duration of 6 months (Ganeshan 1983) with a progressive decline in viability index by germination in vitro at weekly or monthly intervals. Cryogenic temperatures (-196°C) afford much more prolonged duration storage, beyond two years as demonstrated by Japanese workers (Kobayashi et al. 1978). Ganeshan and Alexander (1991) reported cryogenic preservation of lemon pollen for 3.5 years.

Storage of 5°C and -18°C : Storage in liquid nitrogen: Pollen samples are packed either in gelatin capsules or butter paper packets, sealed air tight in polyethylene aluminium laminated pouches and lowered into a canister of a cryoflask. The canister is capped with a perforated lid and plunged slowly in liquid nitrogen contained in the cryoflask. Frequent refilling of the cryoflask with liquid nitrogen at least once every 10 to 15 days, ensures a constant cryogenic temperature. For proper ease of operation, samples to be cryopreserved for a long duration are located at the bottom of the canister and those that are to be retrieved after a short duration, at the top of the canister.

Retrieval and Post-storage Viability Assessment : There are not many critical steps to be followed for retrieval of pollen samples stored at low or sub-zero temperatures. Pollen is removed from the glass vial and directly tested for viability as described above. Pollen samples retrieved from cryogenic temperatures have to be carefully pulled out of the canister with the help of a one foot long blunt forceps, held at ambient temperature for 10-15 minutes prior to a viability test or field pollination.

Pollen samples are viable if:

- Germination profiles in vitro are as good as fresh pollen.
- Germination time taken is not more than that of fresh pollen (pollen vigour).
- Estimated germination is in the range of 60 to 80 per cent of fresh pollen.

Field Pollinations with Stored Pollen for Fertility Assessment

Inventory : Breeder's crossing kit.

Procedure : Stored pollen are required to be tested for fertility under field conditions through controlled pollination. For this, pollen samples retrieved from storage are transported to field locations in an ice box and used in crosses with compatible seed parents in order to obtain an estimate of fertility, fruit and seed set.

Stored pollen is applied to pre-bagged emasculated perfect flowers at peak stigma receptivity with the help of a painting brush, followed by bagging to avoid contamination with stray pollen. For comparative fertility estimates, crosses with fresh pollen must be carried out as far as possible at least initially. If the pollen parent in question is not in flowering, crosses may be repeated with the fresh pollen over the same seed parent within a short period.

Observations to be recorded in field are:

- Number of crosses made with fresh and stored pollen.
- Number of fruits set to maturity.
- Number of seeds per fruit.
- Number of aborted seeds.

Seeds set through stored pollen are germinated and observation of any marker characters transmitted by pollen parent (for example, trifoliate leaf character in F1 population of crosses involving pollen of Poncirus trifoliata), would confirm that the stored pollen was fertile and has transmitted the character conserved to the next generation.

Mango

The species handled for pollen collection include Mangifera indica.

Pollen Collection : Each inflorescence consists of several thousands of staminate and perfect flowers, having a preponderance over staminate flowers. Anthesis is initiated much earlier to the complete development of the inflorescence:

Inventory : Petri dishes, butter paper, forceps, polythene covers (for bagging if necessary) needles, clean razor blade, etc.

Procedure : Pollen collection is usually synchronized with anthesis, between 9 and 11 AM. Anthers dehisce only after the flower is opened. In most Indian cultivars, staminate and/or pistillate flowers have just one or occasionally two anthers producing viable and fertile pollen. These are about 2 mm long and could be seen with the naked eye. In some cultivars, the anthers are pink, turning purple on anthesis, and in some they are dark maroon turning to weak ash-grey colour which is "pollen covered" on the dehisced anther.

Pink or dark maroon anthers are picked out from each flower using a fine pair of forceps and kept on a butter paper in a petri plate. Although it is a laborious process, considerable amount of such anthers are consolidated on a butter paper kept in a petri plate, These anthers on dehiscence turn purple or ash grey. A hand lens can reveal the presence of pollen on the dehisced anthers.

Viability Testing : Spencer and Kennard (1955), and Young (1956) reported in vitro germination of mango pollen. Germinating pollen by the hanging drop method as described for citrus or the modified cellophane method (Alexander and Ganeshan 1989) for recalcitrant pollen resulted in poor viability estimates (Rajasekharan 1996, personal communication). Efforts are under way at IIHR to design a repeatable procedure for germinating mango pollen.

Storage

Inventory: Gelatin capsules, polyethylene aluminium laminated pouches, heat sealer, cryoflask with liquid nitrogen.

Procedure : Mango pollen is required for breeding only once a year and hence long term cryogenic preservation effect was investigated. Anthers laden with pollen were enclosed in gelatin capsules, sealed in polypouches and lowered in canisters and capped as described for citrus. Complete immersion of the samples are ensured by periodical replenishment of the cryogen.

Retrieval and Post-storage Fertility Assessment

Inventory : Field pollination kit.

Procedure : Post-storage fertility of mango pollen is tested by field pollination. At peak stigma receptivity, emasculated flowers of seed parent previously enclosed in polythene bags were pollinated with cryostored pollen by gently touching the receptive stigma with the anther-pollen mixture drawn from the gelatin capsule with the help of fine brush, followed by bagging. Fruit and seed (stone) set could be observed in the next 2 to 4 weeks.

Due to a high inherent fruit drop rate (Mathews and Litz 1992), very few fruits develop among crosses involving cryopreserved pollen as observed in the case with fresh pollen crosses. Cryopreserved mango pollen induced fruit and seed set among intravarietal crosses attempted at IIHR, thereby demonstrating that mango pollen could retain fertility after long term cryopreservation. Further studies on the seed (stone) set with stored pollen is in progress.

Some theoretical and practical considerations of cryogenic preservation of pollen:

1. Techniques of cryogenic storage are new to many plant species, of which the success rate is very high for bicellular pollen material, often collected from tropical regions.
2. Survival of pollen to cryogenic storage and eliciting a high viability response can be used as a model to study the moisture kinetics involved in other plant tissues.
3. The cryogen used *i.e.*, liquid nitrogen though inert, is hazardous by way of causing frost injuries with risk for the personnel using, and those who work in the surrounding area. Training in cryogenic operations, handling procedures, safety precautions, etc., are essential.
4. Places with cryoflasks filled with nitrogen should minimise the risk of gaseous nitrogen build-up. Small volumes of liquid on expansion at ambient temperature, converts into large volumes of gaseous nitrogen, reducing the local oxygen content. This can cause drowsiness, and in extreme cases, asphyxiation.
5. Proper handling equipment, such as cryogloves, long forceps or tongs, face guards if necessary while working for long duration, is essential. It is generally advised not to use glass vials for cryogenic storage, as they can explode due to seepage of liquid nitrogen into the vial during storage.
6. Liquid nitrogen source as far as possible, must be located in proximity and should be easily accessible.

Cryogenic Containers

- Commercially available storage containers with a long static liquid nirogen holding time and with a low evaporation rate are desirable. The canisters designed to hold vials, pouches, etc., in liquid and vapour phase are to be used.

- It is of cardinal importance to ensure transfers or retrievals in and out of the canisters as quickly as possible, so that samples held in storage at -196°C are not allowed to warm up significantly as a result of being relocated temporarily.
- The liquid nitrogen level in storage cryocans will have to be monitored regularly (at least once every week and more frequently if the containers are often opened within this period). Accordingly, the cryogen level has to be replenished to the level recommended by the manufacturer to achieve the required temperature. Keeping the containers in a room temperature controlled between 0º - 10ºC reduces loss of the cryogen to a great extent, thus extending the time between fillings.

Tropical Fruit Germplasm

A considerable variety of tropical fruits are grown in Asia, in particular in South and South-East Asia. The region is a centre of diversity for several important native fruits which include mango (Mangifera indica), Citrus and rambutan (Nephelium lappaceum) (Arora 1994; Kajiura 1990). In addition, more than 200 minor, underutilized native species occur; these are collected in the wild or grown in home gardens. Some of these species have considerable potential, such as breadfruit (Artocarpus altilis) and jackfruit (Artocarpus heterophyllus), which are rich in carbohydrates, proteins, minerals and many vitamins (Negi et al. 1990). Examples of minor fruit crops with high genetic variation in the region are carambola (Averrhoa carambola), longan (Dimocarpus longan) and aonla (Phyllanthus emblica) (Arora 1994).

Just as for other crops, it is important to conserve the genetic material of the various fruits. Human beings are highly dependent on plants for survival. Plants are mainly used for food, but also as sources for building material, fibres, medicines, oils, rubber, perfumes and dyes. The relationships between human being and plants are not static. Because of constantly changing needs in human nutrition, increasing population and industrialisation and trade, a large variety of species is required to meet these needs.

In India and its neighbouring countries, various native fruits, such as aonla, beal fruit (Aegle marmelos), jackfruit, jamun (Syzygium cuminii), karonda (Carissa congesta), Kokum (Garcinia indica) and phalsa (Grewia subinaequalis) are underutilized. Some of these might be important in the near or far future, because of their therapeutic/medicinal and nutritive value as well as their excellent flavour and very attractive appearance. Consumers today are becoming increasingly conscious of the health and nutritional aspects of their food.

Underutilized fruits could play an important role in satisfying demand for nutritious, pleasantly flavoured and attractive natural food of high therapeutic value. Encouraging local people to produce these fruits can help to improve their social and economic welfare. In this way, they can also significantly contribute to the preservation of the environment by stopping uncontrolled harvesting from the wild and assisting in the retention of the various species in their native habitats where they perform best.

Conservation is very important, because many species are becoming extinct and many others are threatened and endangered. The diversity of some fruits is well collected, while for other fruits relatively little has been done yet (Arora 1994). Gaps in collections are found both between species and between regions. This is especially true for both underutilized species and wild crop relatives, where big gaps are noted. Kostermans and Bompard (1993) indicate that Mangifera blommesteinii, M. leschenaultii, M. superba and M. paludosa are in real danger of extinction. High genetic erosion has been noted for jackfruit, Citrus sp. and Litchi chinensis in a survey carried out by the International Centre for Underutilized Crops (ICUC) and IPGRI (Haq 1994).

Furthermore, great genetic diversity is required in order to have a stable ecosystem. In the case of species, which are already used by human beings as crops, it is very important to have a broad genetic base, to improve existing genotypes when necessary. Finally, the diversity of plants in different ecosystems brings a lot of pleasure and inspiration to people with cultural and/or religious significance and the potential for income generation through eco-tourism. Thus, it is important to appreciate the contribution to human welfare and environmental sustainability made by all the three levels of biodiversity: (i) ecosystems, (ii) species, and (iii) genetic diversity (IPGRI 1993).

This paper deals with in situ conservation, particularly as it relates to tropical fruit crops. Two different approaches of in situ conservation are distinguished, namely, genetic reserves and on-farm conservation. Before going into detail on both approaches, major problems and constraints for in situ conservation research are reviewed to give a better understanding on the status of in situ conservation at present. Lastly, various examples, a few on tropical fruits, are presented on how in situ conservation could be applied. These examples can provide a basis for further discussion on how to approach and apply in situ conservation.

Definitions

In situ conservation is defined as the conservation of ecosystems and natural habitats, and the maintenance and recovery of viable populations of species in their natural surroundings and, in the case of domesticated or cultivated species, in the surroundings where they have developed their distinctive properties (UNEP 1992).

Ex situ conservation, on the other hand, is defined as the conservation of components of biological diversity outside their natural habitats, for example, in genebanks (CBFD 1992). Thus, conserved species can be either wild or domesticated and habitats can be either natural ecosystems or agro-ecosystems. In practice, the situation is more complex as different intermediate types exist. For example, IPGRI divides its mandate crops into the following groups: cultivated species and their wild relatives and useful forestry species (Iwanaga 1995). Brush (1991) distinguishes the following four categories of crop germplasm: wild crop relatives, semi-domesticated (weedy) crop relatives, perennial species and landraces of ancestral crop species.

In this paper, the in situ conservation strategy is divided into two general approaches, following the working definitions proposed by Maxted et al. (1997).

1. *Genetic Reserve Conservation:* The location, management and monitoring of genetic diversity in natural wild populations within defined areas designated for active, long-term conservation.
2. *On-farm Conservation:* The sustainable management of genetic diversity of locally developed traditional crop varieties with associated wild and weedy species or forms by farmers within traditional agricultural, horticultural or agri-silvicultural cultivation systems.

In Situ Conservation

Traditional, many crops are conserved as seed in genebanks. However, conservation of tropical fruit crops often takes place in field genebanks due to the recalcitrant nature of the seeds of many of them. Major disadvantages of field genebanks, such as high maintenance costs, the limited amount of genetic variation that can be stored and vulnerability to natural and human disasters have led to efforts to develop in vitro conservation methods.

This conservation can be either under slow growth conditions or through cryopreservation (preservation at very low temperatures in liquid nitrogen). Pollen and DNA conservation are two other techniques, which are being explored and are used on a small scale. All of the above mentioned ex situ conservation techniques have their niche in conserving part of the genetic variation of fruits. However, the conservation of plant material outside its natural habitat brings disadvantages. For example, to conserve material ex situ, it is necessary to remove it from the area in which the plants occur naturally, so interrupting the evolutionary process. Risk of genetic drift increases.

This disadvantage can be resolved by using an in situ conservation strategy. However, this does not mean that in situ conservation is the only or best way of conservation; rather it should be seen and used as an additional and complementary technique (Arora and Paroda 1991; Withers 1991). In situ conservation can be a valuable tool, because it extends the conservation of a species beyond the level of the individual to the habitat or ecosystem in which the species lives.

In addition to preserving the plant material per se, the processes that create new genotypes are also conserved. In other words, evolutionary processes continue letting the plant populations continually adapt to the changing environment. Furthermore, in many case's, it is easier to conserve a viable population in situ than ex situ. This is certainly true of tree species.

In situ conservation might be of great importance for: (i) recalcitrant seeded species, (ii) wild relatives, (iii) wild species, and (iv) species depending on other organisms within the ecosystem, *e.g.*, some plant species have highly specialised breeding systems, which require insect or birds for pollination. For the first group, other solutions which are often very expensive, might result in the conservation of a limited number of accessions. The second and certainly the third group often receive a low priority in collecting, which could be resolved by in situ conservation. For the last group in situ conservation is the only way (Arora and Paroda 1991; Brush 1991; Prance 1997).

Major Problems and Constraints for in Situ Conservation

At present, there is substantial body of information available on the development and establishment of biological reserves. However, there is limited information on the development and establishment of genetic reserves and only very limited knowledge is available on on-farm conservation. Knowledge on the genetic diversity of the target species, agro-ecology and indigenous knowledge are recognised as important elements for on-farm conservation, yet a logical framework is lacking. Iwanaga (1995) recognised and distinguished the following problems and constraints for in situ conservation of crops, their wild relatives, and useful forest and non-forest plant species:

- The absence of a formal institutional framework able to develop and support in situ conservation activities;

- Lack of linkages (both formal and informal) between different groups interested in maintaining or conserving different species and types of plant germplasm in situ.
- Inadequate knowledge of communities who maintain crop or other plant germplasm in situ and an absence of systems for supporting such communities (and frequently, a lack of recognition of their role and interests).
- Low level of awareness by formal-sector agencies of the significance of in situ conservation and the contribution that it can make to biodiversity maintenance and to provision of useful germplasm as a contribution to development.
- Limited number of trained personnel able to guide and lead the development of in situ conservation activities of crop, forestry and other plant species.
- Major limitations in the knowledge base with respect to selection of in situ conservation locations, their monitoring and management.
- A lack of effective analysis of the relationship between development and conservation of the different forms this relationship can take and the impact these will have on the maintenance of the diversity. In the next part of this presentation, existing approaches and frameworks for in situ conservation are described. However, before that, different activities that should be undertaken in order to be able to develop and establish final conservation plan are presented.

Preliminary Information Gathering for in Situ Conservation

Before planning and designing a genetic reserve or an on-farm conservation project, several issues need to be considered. Firstly, the main target and related species should be identified. The choice for these should be objective, based on logical, scientific and economic principles related to value and use of the species (Maxted and Hawkes 1997). Other important factors influencing the choice of a target are threats, which vary from species to species. These threats can be grouped under the headings of overexploitation, habitat destruction or fragmentation, competition from introductions, changes and intensification of land use, environmental pollution and climatic change (Maxted and Hawkes 1997).

A practical approach to select target species could be the use of the genepool concept of Harlan and de Wet (1971). This concept is based on the ease with which species hybridize with each other. Three categories of genepools are proposed:

- *Primary Genepool (GP-1):* This includes the true biological species including all cultivated, wild and weedy forms. Crossing between the various species within this genepool should not cause any problems and the derived hybrids should generally be fertile.
- *Secondary Genepool (GP-2):* This includes those biological species that can be hybridized with the main species, but often result in sterile progeny.
- *Tertiary Genepool (GP-3)*: This includes the species that only could be hybridized with the main species with difficulty, *e.g.*, use of embryo culture, doubling the chromosome number or using bridging species. Resulting hybrids are anomalous, lethal and/or sterile.

The availability of information on different genepools enables the priority setting of target species to be incorporated into the in situ conservation strategy. The general importance of a species and the number selected within a species depends on the ease with which they can be used. So this means that the highest priority should generally go to species in the primary genepool and the lowest to those species in the tertiary genepool (Chapman 1989).

In formulating strategies for the conservation of any crop, it is essential to know its areas of distribution, and to understand the diversity and location of its wild relatives. This knowledge base should consist of both geographical and ecological information. Furthermore, information on the density of the different species is needed.

In order to conserve biological diversity, conservation programmes must be guided by the biology of the species or systems that one seeks to preserve (Karron 1991). The taxonomic status, geographical range, life form, breeding system and seed dispersal mechanism all have significant effects on the levels of genetic diversity maintained by a species (Hamrick et al. 1991). These factors influence the genetic diversity within and between populations of the target species.

Dicots generally have a lower genetic diversity than monocots and gymnosperms. Annuals and short-lived perennials and endemic species normally have a lower genetic diversity than long-lived perennials and wide spread species. Higher diversity is found among cross pollinating species than in self-pollinating species.

The foregoing information indicates that species with limited potential for gene flow have more differentiation between populations, while species with a high potential for gene flow have more differentiation within the population. Gene flow potential is closely correlated with type of pollination, seed dispersal mechanisms, longevity and seed size. Hamrick et al. (1991) concluded that species with high levels of genetic diversity are more likely to have populations with high genetic diversity than those species with low levels of genetic diversity.

In addition to the various technical aspects influencing the choice of target species, and location and size of genetic reserves, socioeconomic and political factors also play a role. Firstly, there are various international agreements and conventions influencing and impacting upon the conservation of plant genetic resources, among which more important are the following:

- The Convention on Biological Diversity (CBD).
- Agenda 21.
- Resolution 5/89 on Farmers Rights.
- FAO Global Plan of Action for the Conservation and Sustainable Utilization of Plant Genetic Resources for Food and Agriculture.

Article 8 of the CBD, a legally binding convention, accepted by the Conference for the Adopting of the Agreed Text of the Convention on Biological Diversity met at UNEP Headquarters in Nairobi, Kenya in 1992, deals with in situ conservation.

Although this is a fairly long article, it can be summarised as follows: the contracting parties should establish a system of protected areas to conserve biological diversity. This should be done in such a way that conservation is ensured and that diversity will be used in a sustainable way. Guidelines for the selection, establishment and management of protected areas should be developed where necessary. These legally binding norms provide a broad basis for both genetic reserves and on-farm conservation.

In situ conservation is also discussed in Chapter 14.G of Agenda 21. Two major objectives of this programme are:

1. to adopt policies and strengthen or establish programmes for in situ conservation;
2. to take care of sustainable use of plant genetic resources for food and agriculture, integrated into strategies and programmes for sustainable agriculture;
3. to take appropriate measures for the fair and equitable sharing of benefits and development in plant breeding between the sources and users of PGR".

Resolution 5/89 on Farmers Rights was added in 1989 to The International Undertaking on Plant Genetic Resources. Resolution on Farmers Rights aims to (Henne 1995):

1. ensure that full benefit to farmers from the exploitation of the diversity that farmers have generated goes to the farmers themselves;
2. support the continuation of the contributions by farmers to the development and maintenance of landraces;
3. attain the purpose of the undertaking. The purpose of the undertaking according to Art. 1 is to ensure that "plant genetic resources, especially species of present or future economic and social importance, are explored, collected, conserved, evaluated, utilized and made available, for plant breeding and other scientific purposes."

Farmers Rights differ from plant breeders right (PBR), in that the objective of Farmers Rights is the conservation and sustainable use of biological resources to serve the community, rather than allowing that biological resources to become a commodity. Furthermore, these rights are a counterpart to intellectual property rights (IPR). In addition to establishing funds to support farmers in protecting and conserving genetic diversity and to inform the public about the importance of conserving these genetic resources, it also guarantees full benefit to farmers of the gains which are realised by others using the plant genetic resources developed by farmers (Henne 1995).

The GPA is a set of recommendations and activities arising from the Report on the State of the World's Plant Genetic Resources, prepared in 1996 on the basis of 150 country reports. The GPA was adopted at the FAO International Technical Conference on Plant Genetic Resources held in Leipzig, Germany in June 1996.

The GPA main objectives are (FAO 1996) to:

- ensure the conservation of PGRFA as the basis of food security.
- promote sustainable use of PGR to foster development and reduce hunger and poverty.
- promote the fair and equitable sharing of the benefits arising from the use of PGR.
- assist countries and institutions to identify priorities for action.
- strengthen existing programmes and enhance institutional capacity.

The GPA refers to both in situ and ex situ conservation. For in situ conservation, four main activities are recommended in this book of the GPA. Firstly, PGRFA should be surveyed and inventoried, in order to be able to develop rational conservation systems and national policies for the use of these resources. Secondly, more support through training, participatory research and a supporting policy environment will be provided for the management and improvement of on-farm conservation.

Thirdly, assistance will be offered to farmers in disaster situations, such as war, civil strife or natural disasters to restore agricultural systems. Lastly, the in situ conservation of wild crop relatives and wild plants for food production will be promoted through improved management of the genetic resources in genetic reserves.

National policies should deal with the national agricultural policy, the national research agenda, genetic conservation programmes, farmer consultation by official agencies, education and incentives for farmers and farming resources (Henne 1995). The 1992 National Integrated Protected Areas System Law of the Philippines is an example in which the legal instrument specifically addresses genetic conservation management.

One of the problems which can be encountered here is that such laws can actually cut across the mandated areas of responsibility of a number of different ministries and departments, such as agriculture and environmental affairs. This makes the application and enforcement of these laws very difficult.

Lastly, it is extremely important to get local communities involved, because they are socially and economically dependent on the environment where the conservation should take place, especially for on-farm conservation. However, even in the case of natural reserves, local communities are often dependent on those areas for example for the collection of food, medicinal plants and fuel wood. The collected material can be either for own use or cash income.

Thus, it is important to explain the purpose of in situ conservation to local communities, to obtain information from them on the area, and to exchange ideas about the selection of a site, using rapid appraisal and participatory appraisal techniques. In other words, indigenous knowledge should be considered as a valuable source of information. Finally, it is very important that communities feel and take responsibility for the entire project. This can make the difference between success and disaster.

Genetic Reserves

A model for how to plan and establish a genetic reserve has been developed by Maxted et al. (1997). This model divides the entire process of establishment and use of genetic reserve into three phases:

1. Reserve planning and establishment.
2. Reserve management and monitoring.
3. Reserve utilisation.

Reserve Planning and Establishment

Using ecogeographic studies, broad areas of genetic diversity are identified. For the establishment of the actual reserve, precise sites have to be defined within these broad areas. Potential sites require a detailed survey based on taxonomic diversity of target species in order to have detailed and up-to-date information on the various sites to facilitate the selection of best sites. This assessment of genetic diversity increasingly involves the use of molecular techniques, such as isozymes or DNA markers. Furthermore, the habitat requirements of the target taxa should be verified and the most suitable sites should contain robust populations of the target taxa.

The sites should hold as much genetic diversity as possible, because the minimum viable population size is often not determined in advance of conservation (Maxted et al. 1997). To prevent total loss of a target species, it is advisable to establish more than one reserve, selecting sites which complement each other. This means that the widest possible range of different ecological and geographic zones in which the target species are found, should be covered.

When considering potential sites for establishing genetic reserves, it would be advantageous to link these to other ongoing conservation interests. In the wider conservation movement, the types of protected area are fairly well understood as defined by the IUCN Commission on National Parks and Protected Areas, and also by various international networks and conventions. As far as possible, genetic reserves should be designed within the general framework set out by these agreements.

Some sort of cost-benefit analysis should be carried out for each site in order to be able to select the most valuable one. It is not easy to measure the benefits, as it is very difficult to estimate the value of a resource especially when it is not of particular interest at this moment, but which might have a value in future. This often means that these figures cannot have more than moral and scientific value. The relative costs of establishing a reserve also affect selection of the site. Obviously, if two sites are of equal genetic value, the one which will be the cheapest to establish should be selected. Due to problems in estimating benefits, the cost-benefit analysis cannot be decisive for the site selection; rather it is an additional information source.

Subsequently, the reserve site should be sustainable in this role for foreseeable future. It must be accessible to the reserve manager and potential users. When selecting a site, the national regulations and policy on plant genetic resources should be kept in mind. It is very important to involve the local community and this should be done at as early a stage as possible.

After the site for a reserve is determined, the reserve itself should be designed. Factors, such as structure, size, whether a single large or multiple smaller sites for the target species, the use of corridors, reserve shape, environmental heterogeneity and potential user communities have to be considered (Maxted et al. 1997).

According to the Man and the Biosphere Programme (UNESCO), a genetic reserve should consist of a core zone surrounded by a buffer zone and outside that, if possible, a transition zone. The core zone should represent a natural undisturbed or minimally disturbed area, representative of the ecosystem. Certain degree of management need to take place and the core species need to be monitored. The buffer zone protects the core zone from edge effects and other factors that might threaten the viability of the target populations in the core. It also allows the reserve to have a more natural boundary. In addition, the buffer zone allows restricted use of the traditionally utilized species by the local communities and it can be used for research, education, demonstration and harmonious interaction with people (Arora and Paroda 1991; Hawkes et al. 1997). The transition zone serves to shield the core zone from general areas of human exploitation.

This zone may be used for the development and research of proper techniques for eco-restoration, limited human settlement, sustainable utilisation and tourism (Hawkes et al. 1997). The size of a genetic reserve should be large enough to ensure self-perpetuation and continuing evolution of the biodiversity within it (Arora and Paroda 1991). This means that the core zone should contain at least 1000 to 5000 potentially breeding individuals to represent a population, the minimum viable population size (Hawkes et al. 1997).

The reserve design can be based either on the use of various small reserves or on one large reserve. For example, is it better to have one large reserve of 15 000 hectares or a network of five, each of 3000 hectares? A large reserve has as advantage that it might cover more habitat types than small reserves. However, the different small reserves can be placed in different environment to enable conserve all extreme ecotypes. Of course, it should be kept in mind that small reserves should be big enough to contain the minimal population size to survive.

The size and number of reserves depend on the characteristics of the target population. Large areas are better for maintaining larger species, while smaller multiple reserves might be appropriate for annual plant species that are often found in dense but restricted stands (Maxted et al. 1997). For the majority of the wild relatives of crop plants, the establishment of multiple small reserves is preferred, because they are generally widely distributed over a range of different ecosystems. The conservation value of multiple small reserves can be increased by the establishment of corridors which facilitate gene flow and migration between the component reserves.

The edge of a reserve should be kept to a minimum to avoid side effects from the outer environment, such as changes in light, temperature, wind, the incidence of fire, introduction of alien species, grazing and deleterious anthropogenic effects. For that reason, a round-shaped reserve has the lowest edge to area ratio. It should also be realised that each disturbance in the reserve, such as roads, dams and fences causes additional edge effects (Maxted et al. 1997).

Lastly, the sustainability of the reserve needs to be considered. In comparison with the maintenance costs of ex situ conservation, the maintenance costs for reserves are high. This is due to the need for active and consistent population monitoring, habitat management and site security, all of which require the commitment of substantial level of resources for a substantial period of time. If material is lost during the in situ conservation process, it could be reintroduced, if stored at another place, either in another reserve or ex situ.

It is also important for sustainability to obtain as much knowledge as possible about future plans for human development for the site selected. Legislation ensuring that once a reserve has been established, it is maintained to assist in securing the sustainability of the reserve (Maxted et al. 1997).

Reserve Management and Monitoring

To be able to manage a reserve in a proper way, a good management plan has to be established. Management plans normally contain information on conservation objectives, site biotic and abiotic dynamics, site history, public interest, factors influencing management, management prescription (what work needs to be carried out and precisely how and when to do it), ecological and genetic survey and monitoring schedule, budget and human resources (Maxted et al. 1997).

In the case of the conservation of certain target species in a reserve, information should be added on the species themselves. This information should be both general (taxonomy, phenology, habitat preference, breeding system and minimum population size), and specific, based on the site (mapping of populations, information on density within the site and interaction of species with the surrounding environment). Furthermore, the plan should contain information on what is an acceptable fluctuation in size of the populations (maximum and minimum values) and when to raise the alarm to take alternative actions to save populations.

A good management plan should also contain detailed information on the appropriate level of intervention through grazing, mowing, burning, planting and harvesting regimes (Maxted et al. 1997). When the management plan is introduced, it should be continuously combined with evaluation, revision and refinement in the light of its practical application. An example of how management can influence a population is seen in the cutting or non-cutting of branches of Myrica trees in Japan. In the protected forest zone along the coastal cliff at Yahatano Village at the Izu Peninsula, large Myrica trees form a part of the upper forest canopy, but young trees are not found inside the forest. On the other hand, in the inland zone forest, where management has been practised to obtain evergreen and deciduous coppice forests, young trees are found at high frequency. This suggest that the availability of more light through cutting has made invasion and growth of Myrica trees possible (Kajiura 1990).

To be able to follow how a population develops over time, it should be monitored regularly. The following activities are part of the monitoring process: defining objectives, identifying key associated taxa and sample quadrat locations, selection of data for quadrat recording, determination of desirable frequency and timing of quadrat recording, accumulation of data sets, statistical analysis and production of recommendations on the management plan (Maxted et al. 1997).

Since it is impossible to record and monitor all species or individual plants, samples are taken and key indicator taxa, mostly including the target species, and key indicator sites are measured. This involves the use of both fixed and random quadrats or transects within the reserve. There are several methods to measure the abundance or diversity within quadrats or transects. The most commonly used is presence or absence with an estimate of density when present. Absolute numbers could also be taken, but that is more time consuming. There is no recommended size for quadrats, but it is the best to keep the same size for all quadrats at every location and time. Furthermore, monitoring should take place in the same season to facilitate comparisons from year to year.

Reserve Utilization

Genetic reserves are not just established for conservation on itself, but to conserve PGR which might go extinct and be needed in the near or distant future. The conserved material is or could be utilized by different groups and for different reasons. Three different groups of users are recognised:

1. local communities.
2. general public.
3. scientific professionals.

The establishment of a reserve will often disrupt the normal activities carried out by local people, such as collecting, hunting or simply visiting the area. To work in the reserve in a sustainable manner, the support of the local community for conservation should be obtained. To avoid disruption to the activities of local communities, wherever possible, sustainable use of the buffer and transition zones by local traditional communities should be encouraged.

The second user group is the general public, who pay for the reserves by taxation. So it is very important to have the support of the general public and make them aware of the importance of conservation. Both ethical and aesthetic reasons are often very well accepted. In addition, the reserves could be used by the general public for example for education and eco-tourism.

The scientific professionals will utilize the conserved material in the same way as material that is conserved ex situ. A disadvantage of the in situ approach might be that the material is not easily accessible for the plant breeder, this can be avoided by characterizing much of the material and disseminate the collated information. The seasonality of the reserve might also limit the access of the reserve, because fruits and seeds can only be collected during the fruiting season of the species. An advantage of in situ genetic reserves for the professionals is that evolutionary studies of the target species can be carried out.

Examples of Genetic Reserves

Myrica rubra is dioecious with its female tree bearing reddish to purple fruits as well as a dye from its bark. It is distributed in coastal areas from southern China to central Japan. Fruit of M. rubra is consumed fresh or brewed into wine or reddish liquor. Ukiyamma, a forest located on a flat lava terrace on the Izu Peninsula in Japan is used as a genetic reserve for M. rubra. In the 19th century, the management of this forest was directly controlled by the Samurai government. Harvesting of the mature fruits of M. rubra in the forest was permitted only for inhabitants. If anyone cut a trunk or even a shoot, the death penalty by decapitation was imposed because M. rubra was an important economic crop at that time. Since then, M. rubra trees have been preserved as common property by means of an agreement by the villagers (Kajiura 1990).

Kostermans and Bompard (1993) recommended that conservation of Mangifera species should be both ex situ and in situ. Although in situ conservation has advantages, since the trees remain in their natural habitat, its big drawback is that the nature reserves in Asia, especially in Indonesia, are not secure. From this example, it is clear that further investigations are needed to explore if in situ conservation is possible or not.

In 1993, a pioneering project on in, situ conservation of wild crop relatives and important forest trees species was started in Turkey. This project aims to provide a national strategy and action plan for the conservation of wild relatives through the development of in situ conservation areas (called management zones) where populations of the most important wild relatives occur. It involves ecogeographic surveys, selection of the sites, development of management plans, and activities to maintain the selected populations, creation of an effective national plan and capacity to carry out the planned programme on a sustained basis (Hodgkin 1996).

At the moment, seven 'Biosphere reserves' have been established in India by the Department of Environment and Forests to preserve biodiversity. There are plans underway for the establishment of another seven. These reserves have been chosen, because they all have a high genetic diversity and also because they represent ecosystems in the different biogeographic regions of the country.

In relation to tropical fruits, it would be more important to have information on the biodiversity held within the core areas of a reserve. Until detailed surveys are carried out in the biosphere reserves, it is rather difficult to have exact information on the diversity of fruit crops. The only information available can be found in the synthesis published by the National Bureau of Plant Genetic Resources (NBPGR) in India (Arora and Paroda 1991).

In the Western Ghats, where the biosphere reserve of Nilgiri Hills is located, jackfruit and mango are known to be distributed. Furthermore, species of Syzygium are found in the Eastern Ghats, where the Gulf of Mannar biosphere reserve is located. Lastly, various Citrus species are found in the northeastern region where the Manas biosphere reserve is located and the Namdapha and Kaziranga biosphere reserves are to be established. Even though information is collected on the presence of fruit crops, no data are available on the density and diversity of each species.

On-farm Conservation

As defined earlier, on-farm conservation of PGR is concerned with crops, and their weedy and wild relatives. On-farm conservation is dynamic and is aimed at maintaining the evolutionary processes that continue to shape genetic diversity. It is based on the recognition that farmers have improved and grown genetic diversity and that this process still continues among many farmers in spite of socio-economic and technical changes.

Farmers play a big role through their selection of plant material which influences the evolutionary process and through their decisions to continue with a certain landrace or not (Bellon 1995).

Landraces are part of agricultural systems. Therefore, on-site conservation cannot be achieved through their isolation in biological reserves. According to Brush (1991) and Bellon (1995), farmers should be encouraged to continue planting their landraces and thereby conserve them. This implies a radically different approach to conservation, one in which farmers play a role equally important as that of scientists and politicians (Bellon 1995). There have been only a few studies aimed at studying on-farm conservation and up to now, no model exists for the development and establishment of on-farm conservation. If on-farm conservation is to be implemented, a fuller understanding of both crop populations on the farming systems that produce them is needed to create active cooperation between farmers and conservationists (Brush 1991, 1995). Bellon (1995) argues that the focus of on-farm conservation should not only be on the varieties a farmer grows, but also on seed flows, variety selection, variety adaptation and seed selection and storage. All elements of the farmers' management influence the whole process of on-farm conservation and these should be studied.

Since scientific back-up for on-farm conservation is lacking, Brush (1991) suggests five general guiding principles that may be used in evaluating proposals and pilot on-farm conservation programmes:

1. Even while focusing on on-farm conservation, it is important to be aware of the complementarity of the different techniques. Ex situ conservation be used as a back-up in case material is lost in situ.
2. Institutional development of in situ conservation strategies should be related to the activities already found in. farming system.
3. The existing institutions and incentives should be reinforced rather than create new ones to provide continuity.
4. Conservation by farmers can be strengthened by agricultural development policies that enhance incentives to continue to maintain landraces. Given the role of farmers in on-farm conservation, meeting development goals such as increased farm income, is critical.
5. Crop germplasm is an international public good and its conservation must be supported through international means.

According to Brush (1991), the main challenge in implementing in situ conservation for crop species is to encourage farmers to continue cultivation of the various landraces. Therefore, he suggests that an institutional framework, an extensive information base and a policy framework are required to develop and establish an on-farm reserve.

IPGRI, based on a number of Brush's recommendations, has developed a global project entitled "Strengthening the Scientific Basis of in Situ Conservation of Agricultural Biodiversity". The project consists of case studies in 9 countries including Nepal and Vietnam. The objectives of this project are to:

1. Enhance and support a framework of knowledge on farmer decision making processes that influence in situ conservation of agricultural crop diversity.
2. Strengthen national institutions in planning and implementation of conservation programmes for agricultural biodiversity.
3. Broaden the use of agricultural biodiversity and to involve farming communities in its conservation.

As a first step, a basic data set will be collected through interviews on farmer decision making in the selection and maintenance of landraces to provide measurable indices of genetic diversity of the landraces. In addition, morphological and agricultural information relating to characteristics that the farmer does not actively select for are also being gathered. Finally, biochemical and molecular markers will be used.

Once a basic data set is established, key factors that affect farmer decision making will be identified. A wide range of environmental biological, socioeconomic and/or cultural factors will be included. To obtain all the information described above, IPGRI collaborates closely with national programmes. To support national capacity building, IPGRI will also provide training on in situ conservation in the fields of conservation biology, populations genetics, ecology, etc.

Broadening the use of agricultural biodiversity will be based on the following activities:

- Decentralised farmer managed selection and crop breeding.

- Market and non-market incentives, resource tenure and promotion of local plant uses.
- National policies which enhance maintenance.

Participation of farming communities and the informal sector is very important in these activities. Lastly, it is important to collect information about the sustainability of on-farm conservation. This could be done on-basis of following questions (Brush 1995).

- Will farmers continue to maintain landraces even if land distribution systems change?
- In many places, landraces are still conserved, however, in a much smaller scale (Brush 1991, 1995). How will this reduction affect crop evolution?
- What influence has human population growth on on-farm conservation sites?

All of the limited and fragmented information available on on-farm conservation is related to main food crops, such as cereals and legumes. Very limited information exists on the on-farm conservation of tropical fruits. However, two on-going IPGRI projects in which tropical fruits play a role are described in the following examples.

Examples of on-farm Conservation

Part of an on-going project on "Diversity Conservation and Sustainable Use of Genetic Resources of Fruit Trees Native to Tropical America" deals with on-farm conservation of Sapotaceae handled by IPGRI Americas office. Sapotaceae, large trees producing edible fruits, are cultivated throughout Central and South America in home gardens producing fruit for domestic consumption and additional income. Since Sapotaceae have recalcitrant seeds, in situ conservation will be an important component of any conservation strategy.

The following activities are undertaken to promote on-farm conservation. Firstly, monitoring of genetic diversity and genetic erosion are studied. Then information is gathered on farmers' decision-making. To promote the use and on-farm conservation of Sapotaceae, agronomic and post-harvest techniques are being developed, advanced materials selected and socio-economic studies on production including marketing conducted.

A pilot study on home gardens is being carried out in Vietnam to evaluate the contribution of home gardens to in situ conservation (Hodel 1997). Since several tropical fruits are grown in home gardens, this on-going study might offer useful information that could be useful to other projects on on-farm conservation of tropical fruits.

Conclusions

There is a high recognition of genetic reserves and on-farm conservation at present. Therefore, these techniques should be developed further, especially with regard to measuring genetic diversity over time, integrating the various sciences relevant to in situ conservation, developing generic models, and developing local, national and international policies. In situ conservation appears to be promising for tropical fruit crops. Results from different on-going IPGRI projects, especially those relating to tropical fruits, will give better insight as to how to develop and establish future in situ projects.

Complementary Conservation

Crop genetic resources collections are assemblies of genotypes or populations representatives of cultivars, genetic stocks, and related wild and weedy species, which are maintained in the form of plants, seeds, tissue cultures, etc. (Frankel and Soule 1981). Functionally, such plant genetic resources (PGR) include landraces, advanced/improved cultivars, and wild and weedy relatives of crop plants. Landraces may contain co-adapted gene complexes that have evolved over decades (Harlan 1992) and are the most important of the PGR. Advanced cultivars, genetic stocks and wild relatives of crop plants have a part to play in future crop improvement and, therefore, need to be preserved (Frankel 1990).

The great wealth of genetic diversity remaining in the genepools retains vast potential for present and future use. Broadly speaking, PGR are non-renewable and it is essential that we should be concerned with conservation, be it at species level, genepool level or at the ecosystem level. The limitations of a narrow genetic base of many present day varieties have been stressed many times. Genetic diversity is a defence against the genetic vulnerability, which has been built into the genetic structure of traditional cultivars (Anon 1973; Brown 1983). Countries which still have a significant amount of genetic diversity and species diversity have a responsibility to themselves as well as to the world at large to conserve it and make it available for use (Ramanatha Rao 1994).

Conservation Approaches and Methods

It is well known that there are two approaches to conservation of PGR - ex situ and in situ. Ex situ conservation approach generally comprises the following methods: seed storage, field genebanks, in vitro storage, pollen storage, DNA storage and botanical gardens. Conservation of plant diversity using reserves/protected areas, on-farm and home gardens is considered as in situ conservation approach. We will now look at each method briefly.

Ex Situ Conservation

Conservation of Seeds : In the past, many collections were maintained without the help of storage which would extend the viability of seeds. Due to this, the conserved accessions had to be regenerated very frequently leading to loss of genetic diversity in genebanks (Frankel and Hawkes 1975). Generally speaking, seed storage conservation method is an efficient and reproducible technique. The seeds are dried to a lower moisture content and stored at a low temperature. This method is almost universally applied to the orthodox seed species.

However, the procedures followed in conservation, maintenance and storage present many problems. In maintaining genetic purity of the conserved accessions, problems arise due to differential survival in storage, selection during regeneration, outcrossing with other entries and genetic drift (Allard 1970). Good storage conditions coupled with proper grow-outs are expected to reduce the effects of such problems (Rao 1980). It is well recognized that seeds should be maintained under conditions in which the life processes in seeds are minimized so that they can be stored for a number of years with little loss in genetic diversity, genetic integrity and viability. In general, limits can be set for loss of viability as indicator of the genetic shifts in the stored seed material.

Due to the extended life span of seeds stored under optimum conditions (cool and dry), it will not be necessary to regenerate the seeds at frequent intervals. Seeds of most common species can be maintained in this way for a number of years. Reduced frequency of regeneration results in more cost-effective maintenance of germplasm. More importantly, it minimizes the genetic erosion resulting from genetic drift in small populations that may be grown for regeneration; genetic shifts resulting from natural selection when material is grown out in the field (especially when grown under completely different conditions from that of the original location) and the risk of human error (Frankel and Hawkes 1975; Ramanatha Rao 1991). Guidelines for proper handling and storage of seeds of many different crop species are now available (FAO/IPGRI 1994; IBPGR 1982, 1985a, 1985b, 1985c, 1985d).

Work is also in progress on alternative methods of storage of seeds such as the maintenance of seeds imbibed storage, storage of seeds in liquid nitrogen at temperatures below-196°C and storage of ultradry seeds. As opposed to common seeds, generally called 'orthodox' seeds which can be dried to very low levels of seed moisture content (below 7%), there are a number of species whose seeds cannot be dried to low levels for optimum storage. Such seeds have been referred to as 'recalcitrant' (Roberts and King 1986) and imbibed storage (at higher levels of seed moisture) may be of considerable importance. Efforts are also underway to undarstand the genetic basis of recalcitrance which will be necessary to deal with this problem. Very low temperature storage using liquid nitrogen, called cryopreservation, also appears to be promising with a more extended life span than seeds stored in currently what is described as long term storage (-20°C). Another area in which considerable work is required is on storage of ultradry seeds (dried to seed moisture content of 2-5%) at room temperature conditions. It is now known that such low levels of seed moisture content would help greatly in extending the seed longevity under ambient conditions, when stored in hermetically sealed containers (Ellis and Roberts 1991; Zhou et al. 1995). However, more research will be necessary before ultradry seed technology can be adopted.

Conservation of Plants

Many important varieties of field, horticultural and forestry species are either difficult or impossible to conserve as seeds (*i.e.*, no seeds are formed or if formed, they are recalcitrant) or reproduce vegetatively. Hence, they are conserved in field genebanks (FGB). FGBs may run a risk of being damaged by natural calamities, infection, neglect or abuse. Ex situ conservation of tree species using FGBs requires a substantial number of individual genotypes to be an effective conservation measure.

Thus, FGBs require more space, especially for large plants such as tree species, and they may be relatively expensive to maintain depending upon the location and the complexity of alternative techniques available. However, FBGs provide easy and ready access to conserved material for research as well as for use. For a number of plant species, the alternative methods are not fully developed yet so that they can be effectively used (Ramanatha Rao and Riley 1995; Ramanatha Rao et al. 1996). Thus, it is clear that establishment of FGBs will play a major role in any conservation strategy for PGR. It is one of the options of a complementary strategy for the conservation of germplasm of many plant species. At the same time, efforts to develop and refine other methods, such as in vitro conservation and on-farm conservation must continue.

Conservation of Tissues/cells

Another method of ex situ conservation of germplasm is through the use of tissue cultures. There are a number of important plant species which cannot be conserved as seeds and present different problems. Generally, conservation of such material as plants in the field requires greater space, labour and costs. They also run the risk of being damaged by natural calamities and virus infections as noted earlier. Hence, the conservation of tuber, root, shrub and tree species becomes very difficult indeed. Several techniques to conserve such vegetatively propagated species have recently been developed and some of them are undergoing rigorous testing. Possibilities now exist to conserve PGR as tissue cultures. For some species, the in vitro conservation may be the only option available.

Though tissue culture offers great potential for conservation of germplasm of vegetatively propagated material, two things have been of major technical hindrance to it. Firstly, the genetic instability of the material conserved as tissue culture due to somaclonal variation at the time of regeneration of the tissue into seedlings. Secondly, the length of storage as tissue has been limited.

Significant work is being done on both aspects and for some species, tissue culture maintenance is relevant due to improved techniques resulting in low levels of somaclonal variation. Work on cryopreservation of tissue culture so that these can be preserved for long periods is also making rapid progress. Once these techniques are further refined, their large-scale adoption will be possible (Engelmann and Ramanatha Rao 1996; Ramanatha Rao and Riley 1994a, 1994b; Simpson and Withers 1986; Withers 1993).

Conservation of Pollen

Pollen storage was mainly developed as a tool for controlled pollination of asynchronous flowering genotypes, especially in fruit tree species (Alexander and Ganeshan 1993). In addition, pollen storage has also been considered as an emerging technology for genetic conservation (Harrington 1970; Roberts 1975; Withers 1991). Even if it may not be considered to be a viable method for meaningful genetic conservation of genotypes, cryopreservation is likely to be more successful than other storage techniques routinely employed for pollen (*e.g.*, under organic, desiccation freeze drying, low temperature) in facilitating hybridization when flowering is asynchronous or for use in next season. Thus, it can help in better utilization of available genetic resources. Pollen can be easily collected and cryopreserved in large quantities in relatively small space. In addition, exchange of germplasm through pollen poses fewer quarantine problems compared with seed or other propagules.

In recent years, cryopreservation techniques have been developed for pollen of an increasing number of species (Bhat and Seetharam 1993; Towill 1985) and cryobanks of pollen have been established for fruit tree species in several countries (Alexander and Ganeshan 1993).

DMA Storage

Storage of DNA is, in principle, simple to carry out and widely applicable. The storage of DNA seems to be relatively easy and cheap. The progress in genetic engineering has resulted in breaking down the species and genus barriers for transferring genes (NRC 1993). Transgenic plants have been produced with genes transferred from viruses, bacteria, fungi and even mice. Such efforts have lead to the establishment of DNA libraries, which store total genomic information of germplasm in the form of DNA libraries (Mattick et al. 1992). However, strategies and procedures have to be developed on how to use the material stored in the form of DNA. Therefore, the role and value of this method for PGR conservation is not completely clear yet.

Botanical Gardens

There are about 1500 botanic garden and arboreta in the world (WWF-IUN-BGCS 1989). The objective of most of the gardens are said to be to: (i) maintain essential ecological processes and life support systems, (ii) preserve genetic diversity, and (iii) ensure sustainable utilization of species and ecosystem. However, the botanical gardens may play a limited role in the context of conservation and propagation, and probably a greater role in public awareness and education.

Botanical gardens may mainly be used to display a great number of different and exotic species. As the numbers that can be maintained in this manner is limited, it can not reflect or conserve genetic diversity. There is a possibility that a few well managed gardens can lay emphasis on conservation of certain groups of species as living collections (*i.e.* field genebanks).

In Situ Conservation

This type of conservation is dynamic as opposed to the semi-static nature of ex situ conservation. One of the reasons given for choosing in situ conservation over ex situ is the need to maintain the evolutionary potential of species and populations (Frankel 1970; Frankel and Soule 1981; Ledig 1988, 1992). However, given the fact that human activities can cause habitat destruction and loss of biodiversity in some cases, and the maintenance of biodiversity in other cases, the need to complement it with ex situ conservation efforts is well recognized.

In general, research and monitoring is needed at three levels for successful in situ conservation: the assay of genetic variation represented within a target species in a particular area (ideally by studies of intraspecific morphological and molecular variation and the diversity as recognized by local users including farmers); regular inventory of species numbers; and observation of general ecological condition and habitat alteration including farming systems (Berg 1996).

Biosphere Eeserves/protected Areas

In general, the biodiversity at the species and ecosystem level can only be conserved through in situ conservation (McNeely 1996). Various types of protected or semi-protected areas that are identified to be rich in diversity of ecosystems and/or species are used in this method. Conservation of wild species crop relatives in genetic reserves involves the location, designation, management and monitoring of genetic diversity in a particular, natural location (Maxted et al. 1997). However, it must be remembered that genetic reserves are often not very accessible for use. Additionally, if the monitoring and management may not be optimal due to difficult conditions under which these need to be performed. For the same reason, characterization and evaluation will be limited. The reserves are also vulnerable to natural and human-made disasters.

On-farm Conservation

In situ conservation of agrobiodiversity or on-farm conservation involves the maintenance of traditional crop cultivars (landraces) or farming systems by farmers within traditional agricultural systems (Hodgkin et al. 1993; Ramanatha Rao 1997). Traditional farmers use landraces, which are developed by the farmer and strongly adapted to the local environment (Harlan 1992). This method of conservation has been gaining importance in recent years, though farmers have used it for centuries.

In the case of agrobiodiversity, the effects of growers-practices are of paramount importance. In almost all such areas, there is presently little information available on the status of the genetic diversity. It is now possible to monitor and estimate genetic diversity using molecular markers (Hodgkin and Debouck 1992; Ramanatha Rao et al. 1997; Ramanatha Rao and Riley 1994a, 1994b). However, the limited resources available for such work makes such a proposal difficult to implement. Measurement of genetic diversity in most perennial tree species depends largely on morphometric traits, but increasing use of molecular markers can assist in better understanding of the structure of genetic diversity both at a specific site and across regions.

Systematic documentation of farmers' knowledge of diversity and uses is needed. Sustainable in situ conservation will require community participation, control of land rights in local communities, education, extension and development of environmental awareness. Of equal importance is the principle that any in situ conservation programme must benefit the local communities. Management by local communities can often be developed to effectively link conservation and use (McNeely 1994, 1996). It is important to consider indigenous knowledge, peoples participation and cooperation between local people, researcher and conservationists and non-governmental organizations (NGOs). Additionally, it is important to consider the establishment of areas of intensive management or high yielding plantations for long term sustainability of any in situ conservation programme. Conservation activities by commercial and private agencies can also be promoted as these groups have the capacity to fund such activities. Since it will be necessary to foster sustained conservation and use of resources to derive long-term benefits from the exploitation of the resources, we believe that the commercial sector and private agencies will be interested in activities mentioned above. This can lead to much wanted linkages among public, community and private sectors in plant genetic resources conservation (Riley 1995). The methods of management and benefits to local communities in maintaining and using this diversity must be considered while implementing an in situ (or on-farm) conservation programme.

Home Gardens

Home garden conservation is very similar to on-farm conservation, however, the scale is much smaller. In most rural situations, home gardens tend to contain a wide spectrum of species, such as vegetables, fruits, medicinal plants and spices, than on-farm plots. As it is akin to on-farm conservation, the dynamic nature of this conservation technique has same advantages. Home garden, as a single unit, has very little value in terms of conservation, but a community of them in a given area may contribute significantly to the conservation and direct use of genetic diversity. Most of such diversity could be somewhat unique/rare as the people tend to grow unique materials in their gardens and also underutilized species or even un-domesticated species. However, the system is vulnerable to changes in management practices. Home gardens are also known to be testing grounds for farmer-home gardener as well as a location for testing out some of the wild and semi-wild species. Thus, in rural areas, the home gardens will continue to play a role in genetic diversity conservation as well as development.

Complementary Conservation Strategy

It is important to emphasize that these two main approaches to conservation of PGR: ex situ and in situ are complementary in nature. Conserving a genepool should employ a combination of methods, from nature reserves to genebanks as no single method can conserve all the diversity. The appropriate balance between different methods employed depends on factors such as the biological characteristics of the genepool, infrastructure and human resources, number of accessions in a given collection and its geographic site, and the intended use of the conserved germplasm. For any given genepool, the extent of a particular method used may defer from that used in another genepool (IPGRI 1993).

Genepool

Focusing conservation on genepool basis appears to be the best approach, since it will help us to conserve maximum genetic diversity of any group of crop or forestry plant species (IPGRI 1993). So for effective collecting and conservation, it is important to determine the genepool of species. Most commonly three types of genepools have been distinguished (Harlan and de Wet 1971):

- *Primary Genepool:* This corresponds with the botanical species. It consists of cultivated species and its weedy and wild relatives, which can cross with ease among themselves and produce fertile F1s.
- *Secondary Genepool:* This consists of phylogenetically related species, which can be crossed with the species of interest, and may produce partly fertile, but weak F1s.
- *Tertiary Genepool:* As the name indicates, it consist of species that are phylogenetically related and normally do not intermate. However, they may be crossed using complicated techniques, producing weak and often sterile F1s.

Except monotypic species, all others are expected to have relatives that fall into this classification. With the advent of modern biotechnological techniques that can transcend species/ genus barriers, the definition of genepools is slightly blurred, but this concept still helps us to group the target species and its related species into understandable divisions. Once the target species and the genepools are determined, steps in developing complementary conservation strategy can be taken.

To formulate a conservation strategy, the conservationist need to address a number of questions in relation to the genepool. In the beginning, we have noted that any complementary conservation strategy depends on factors, such as the biological characteristics of the genepool, infrastructure and human resources, number of accessions in a given collection and its geographic site and the intended use of the conserved germplasm. These factors can be used to formulate the following questions. These are not exhaustive, but cover most of the ground.

Biological characteristics Genetic diversity:

- What is known about the genetic diversity within the genepool, at each level?
- What special traits or unique diversity should one be looking for?

Life Cycle:

- Does the genepool include only short lived or long lived or a mixed group?
- What is the age of the plant when it becomes useful for either breeding or for other purposes?

Reproductive Biology :

- Do all species under consideration reproduce sexually or asexually, or a combination of the two?
- If reproduction is sexual, then is it a selfing or outcrossing species?
- What is the level of outcrossing?
- Are there any other biological traits that will interfere with true seed production?

Storage Characteristics:

- Are species under consideration orthodox or recalcitrant seed species?
- Should the storage be for short, medium or long-term?

Location and Facilities :

- Where is the germplasm located?
- Can all the species in different genepools be maintained or grown in the same location?

Infrastructure:

- How accessible is the area of collection?
- How accessible has the material to be and for whom?

- What facilities for different types of conservation are available?

Human Resources:

- What is the level of training of the staff?
- Can the level be upgraded periodically?

Objectives of Collection:

- Is the aim to conserve genes or genotypes?
- Will it completely stop evolutionary processes?

Improvement Strategy:

- Is the collection linked to the user community?
- What kind of breeding methods are used?

Others Factors—Legal Issues:

- What legal arrangements exist for transfer and access to genetic material?
- What are the legal issues that govern the acquisition of land and other related facilities?

Importance of the Crop:

- How important is the crop?
- Is it important enough for the local people to participate in situ conservation effort?

Model Complementary Conservation Strategy

If most of the above questions are answered, then an appropriate conservation method can be determined. A combination of methods, from both the approaches, which are complementary, should be applied. Such a combination should balance amount of germplasm that can be conserved using each method so as to provide us with a strategy that optimally conserves maximum diversity within the target genepool.

For example, if one considers the conservation of Citrus genepool, at the simplest level, one has Citrus species that have orthodox seeds or intermediate seeds or recalcitrant seeds. All other things being equal, one could say that the Citrus species that produce orthodox seed are conserved in seed genebank (However, this can be complicated by many other considerations, for example, objective of conservation, polyembryony, etc.). Species that produce recalcitrant seeds, depending on the level and stability of the technology available may be conserved as tissue (slow growth or cryopreservation) or in field genebanks. If the decision is to use former, the next question to answer is what is the gestation period for in vitro conserved material to become useful. In this fashion, it will be possible to go through the list of questions and determine what part of the genepool will be conserved using which method, thus establishing a complementary conservation strategy for Citrus. As one can see, there is a need for complete information to make the appropriate choices. Personally, I think that attempts to develop models at species group level may be futile as the number and type of parameters considered for any one genepool may vary greatly, and these will make any model effectively unworkable.

Concluding Remarks

No single method can help us to conserve most of the genetic diversity in any given crop or genepool. What we need is a complementary conservation strategy which make, use of optimally all techniques/methods available (and leave place for any that need to be developed or refined). It is a process of making decision about the conservation of a genepool. Such a strategy should be able to meet most of the needs in PGR conservation.

It should be pragmatic and at the same time critical, especially in determining the methods to be used. It does not advocate methods, simply because the method is available, but only because it is the most appropriate one under the given conditions. A good complementary conservation strategy does not categorize crops or species into definitive classes. It is dynamic, and lends itself to meet the challenges of changes that are occurring in the field of genetic resources as it is open to new technologies and new needs.

8

Germplasm on Fruit Crops

India is an important centre of origin and diversity of many horticultural crops including fruit crops like mango, citrus and banana. The sub-continent (area 3 276 141 km^2) extends from 8° 04' - 37° 06' North latitude to 68° 07' - 97° 25' East longitude, and has tropical, subtropical and temperate climate.

Therefore, a variety of fruits originating in tropics, subtropics and temperate regions of the world have been introduced in India and many of them are commercially grown in the country. The Indian Institute of Horticultural Research (IIHR) is a premier research institute for horticulture established at Bangalore on 1st February 1968.

It has four regional Central Horticultural Experiment Stations at Ranchi (Jharkhand), Godhra (Gujarat), Chettalli (Karnataka) and Bhubaneshwar (Orissa). The main institute and its regional stations have the mandate of ex situ conservation of genetic variability of both indigenous and exotic germplasm of horticultural crops, namely, fruits, vegetables, ornamental, medicinal and aromatic crops.

The plant breeders involved in crop improvement of specific crops look after maintenance of field germplasm collection and undertake characterization, evaluation and documentation of plant germplasm collections and their distribution to user scientists. The Section of Plant Genetic Resources has the mandate to carry out and coordinate institution's activities on germplasm collection and conservation of horticultural crops. This paper discusses this aspect with emphasis on fruit crops.

Germplasm Collection and Conservation

The number of accessions collected and conserved ex situ in field genebanks at IIHR and regional stations are as follows (Tables).

Germplasm of Fruit Crops Available at IIHR, Bangalore

Sl.No.	Crop	No. of Accessions
1.	Mango	170
2.	Banana	290
3.	Grape	579
4.	Citrus	82
5.	Guava	24
6.	Pomegranate	52
7.	Annona	16
8.	Fig	19
9.	Jujube/Ber	38
10.	Others	15

Germplasm Collections at CHES, Ranchi

Sl. No.	Crop	No. of Accessions
1.	Mango	120
2.	Guava	19
3.	Litchi	17
4.	Banana	112
5.	Grape	28
6.	Jackfruit	44
7.	Pomegranate	12
8.	Bael	8
9.	Annona	7
10.	Jujube/Ber	31
11.	Indian gooseberry/Aonla	5
12.	Others	8

Germplasm Collections at CHES, Chettalli

Sl. No.	Crop	No. of Accessions
1.	Mandarins, mandarin hybrids and tangeloes	46
2.	Rangpur lime	16
3.	Pummelo, grape fruit, citron	11
4.	Lemon and lime	9
5.	Rough lemon, Karna Khatta	11
6.	Trifoliate oranges	11
7.	Sweet oranges	10
8.	Citranges	5
9.	Citrumelo, citrimon, kumquat	6
10.	Sour oranges	6
11.	Citrus species, related genera	14

Germplasm Collection at CHES, Godhra

Sl. No.	Crop	No. of Accessions
1.	Jujube/Ber	40
2.	Pomegranate	9
3.	Indian gooseberry/Aonla	10

Introduction and Exchange

Introduction and exchange of germplasm of fruit crops is faced with several difficulties. Although seed is the most ideal and convenient medium for import of germplasm of vegetables, cereals, oilseeds and other annual crops, it cannot be used in majority of fruit crops excepting papaya and macadamia. In few other crops like citrus, introduction through seed is restricted for raising rootstock species/varieties. Import of germplasm through bud-sticks and cuttings is successful only in deciduous crops, such as grape, fig, pomegranate, apple, etc. In such cases also, the cuttings are received mostly in half dead conditions as the material is first received by NBPGR, New Delhi for quarantine clearance and then sent to the Institute. Therefore, the percentage of successful sprouting and establishment is very poor. Practically very little introduction/ import of clones of evergreen tropical fruit crops like mango and avocado has taken place in past two decades. If import of grafted plants is not feasible, we may have to resort to introduction of some variability through seed as the second best alternative.

Seed Conservation

Long term storage of seeds of fruit crops in genebank at -20°C in laminated aluminium foil pouches after drying them to around 6% moisture content is not very effective for following reasons:

(1) With exception of papaya and macadamia, no other varieties of fruit crops produce true to type from seeds.

(2) Seeds of majority of citrus species, *e.g.*, mango, jackfruit, jamun, durian, which are used for raising rootstocks are recalcitrant. Vegetatively propagated clonal materials and recalcitrant species are required to be maintained under field conditions. The institute has recently started strong programme on in vitro conservation and cryopreservation of tissue/cell cultures, embryos and gametes.

Germplasm can be conserved in several ways, however, conservation in the form of seeds has many advantages as it is simple to use, easy to handle, practicable, inexpensive and capable of maintaining genetic stability during storage. It is desirable to conserve most of the perennial species like fruit crops in field as fruit repository. Further, this could be supplemented with possible seed storage for short term conservation.

Storage Behaviour of Seeds

Long term seed storage serves as a safe and relatively inexpensive method of plant genetic resources conservation (Hong et al. 1996). Seed longevity varies within the species and in some cases within the cultivars. Several preharvest factors determine the longevity like cumulative effect of environment during seed maturation, harvesting, drying and the pre-storage environment, time of seed harvesting, duration of drying and the subsequent period before seed is placed in store (Hanson 1984). Roberts (1973) classifies seeds into two groups according to their storage behaviour; orthodox and recalcitrant. Of late, a third category, viz., intermediate between the orthodox and recalcitrant has been identified.

Orthodox seeds withstand to drying especially to low levels of moisture content and over a wide range of environments, their longevity increases with decrease in seed moisture and storage temperatures. Orthodox seeds do not tolerate desiccation at all stages of their development and maturation.

Recalcitrant seeds cannot be dried without damage. Fresh recalcitrant seeds lose viability slightly as they are dried, subsequently viability drops considerably at "critical moisture content" Further, drying results in complete death of the seeds. Critical moisture content levels vary among species, cultivars and seed lots (King and Roberts 1979; Chin 1988).

Differences in moisture content have been observed among tissues within particular recalcitrant seeds. The storage tissue of recalcitrant seeds are always at a lower moisture content than the embryonic axes except durian and jackfruit (Grabe 1989).

Desiccation of excised embryos and embryonic axes has considerable potential for the in vitro conservation of recalcitrant embryos, since embryos are capable of surviving desiccation to lower moisture contents than the whole seeds (Chin 1988). In certain cases, fast drying allows intact recalcitrant seeds to survive desiccation to lower moisture contents than slow drying (Pritchard 1991).

Fu et al. (1993) reported that drying excised embryonic axes by silica gel or an aseptic air current allows excised embryonic axes to survive desiccation to a lower value than that achieved by the vacuum method.

Although vacuum drying method provides more rapid drying, no excised embryonic axes of jackfruit survived desiccation to 44% moisture content, while the excised embryonic axes dried with an aseptic air flow and silica gel tolerated desiccation to 26 and 16% moisture content respectively.

Storage Methods for Recalcitrant Seeds

Moist Storage

There is no satisfactory long term method for maintaining viability of intact recalcitrant seeds. This is due to their non-drying and inablility to store at sub-zero temperatures, because seeds would then be killed by freezing injury due to ice formation. Some tropical recalcitrant seeds are also damaged by chilling injury at temperature of 10-15°C. The longevity of recalcitrant seeds of tropical crops is generally short from a few weeks to a few months (King and Roberts 1980). The seed longevity of species belonging to temperate environment can be maintained for much longer period. Thus, species with recalcitrant seeds can be sub-divided into those of tropical origin and those adapted to temperate climates (temperate latitudes or high altitudes in the tropics), and the latter can be stored at cooler temperatures and for longer period.

Cryopreservation of Excised Embryos

The excised embryos of durian and jackfruit survive desiccation up to 30-33% (Hor et al. 1990). Methods of drying and rate of drying are important factors influencing the desiccation tolerance of excised embryos or embryonic axes. However, it is still in experimental stages and need some more refinement of work in this regard.

Intermediate Seed Storage Behaviour

Seeds of certain species exhibit storage behaviour between the orthodox and recalcitrant categories *e.g.* papaya and several Citrus species (Hong and Ellis 1995). The essential feature of intermediate seed storage behaviour is that the negative relation between seed longevity in air dry storage moisture content is reversed at values below those in equilibrium (at 20°C) with about 40-50% RH.

Most of the seeds exhibit desiccation injury to the seed immediately after desiccation to relatively low moisture content, about 7-12% moisture content, depending upon the species. Another characteristic of intermediate seeds of tropical origin is the longevity of dry seeds (7-10% mc) which reduces with reduction in storage temperature below 10°C (Hong and Ellis 1992). Thus, it is desirable to preserve such seeds at 10°C. Intermediate seeds of species of tropical origin die more rapidly when the temperature is lowered below about 10°C (Hong and Ellis 1992). Thus, it is desirable to preserve such seeds at 10°C. Intermediate seeds of species of tropical origin die more rapidly when the temperature is lowered below about 10°C and seeds of a few species are killed as the temperature is reduced to below 0°C.

Processing in Fruit Crops

Seed Extraction

In general, fruits or seeds for storage should be harvested at fully matured stage, otherwise it affects the seed longevity. Tree seeds should be collected from a single plant if not sufficient, then seeds from neighbouring plants with same age be collected. It is safer to carry the fruits than seeds. Moist extracted seeds may germinate or begin to germinate during shipment.

Also extracted seeds may lose moisture rapidly and damage the seed quality especially in recalcitrant types and may rot. It is desirable to store the fruit temporarily when delay is expected for seed storage. A relatively small quantity of seeds are required for long term storage. It is desirable to remove seeds manually, followed by soaking in water and fermentation and chemical treatments.

Seed Drying

Moisture content of seed is most important factor in determining the seed longevity. After removal of seeds, simple washing is sufficient to remove any substance adhering to the seeds. Further, seeds are to be dried based on their storage behaviour.

It is desirable to dry seeds to 5-7% moisture in orthodox group. Seeds of tropical species should be dried at cool temperature of about 15°C. Delay in drying or drying at higher temperature considerably affect the viability in orthodox seeds especially those rich in oils. Silica gel is an effective desiccant commonly used for seed drying.

Seed moisture is determined as per the guidelines prescribed by International Seed Testing Association (ISTA). For large seeds, moisture content determination of about 20-25 individuals seeds is recommended. Separate determination of the moisture content of embryo or embryonic axes, cotyledons or endosperm and the seed covering structure may also be helpful in interpreting the results.

Seed Packing

Seed storage in suitable containers will prevent the direct contact of seeds with storage environment and also protect from pests and diseases. Containers like paper, cloth, polyethylene bags, glass bottle, cans and aluminium foil laminated pouches are being used as containers. The choice of containers depends on kind and quantity of seeds, duration and condition of storage. Paper and cloth bags are cheap and can be used for short term storage. Aluminium cans and foil laminated pouches are effective for retaining the viability for longer period.

Storage Environments

Storage temperatures, humidity and oxygen are important factors associated with storage of seeds. The metabolic activity and incidence of pests and diseases increase at high seed moisture and high storage temperatures resulting in loss of viability. Thus, low moisture at low temperature is effective in maintaining seed viability especially for orthodox type of seeds. Such seeds are being stored in moisture proof containers at -20°C for very long period. However, seeds can be stored at 5 to 10°C for medium term storage. While undried seeds of recalcitrant species are to be packed in semi-moisture resistant containers along with measures to prevent attack of saprophytic fungi and stored at medium temperature (10-15°C), to keep seed alive for shorter period (Doijode 1995).

Seed Storage in Certain Fruit Crops

Mango

Mango seeds are protected by hard stony endocarp. Seed is ex-albuminous having two fleshy cotyledons with testa and tegmen represented by two papery layers. Polyembryony is observed in certain cultivars. Seed exhibits recalcitrant type of storage behaviour wherein seed loses the viability on drying after extraction and undried seeds cannot be preserved long at low temperatures.

Reduction of seed moisture to below 25% in Alphonso and 32% in Totapuri lowered the germination percentage and seedling vigour (Doijode 1990). Patil et al. (1986) obtained 40% of germination in Alphonso stones after 90 days of storage at 25°C in polyethylene bags with charcoal powder. It declined from 75 to 12% after 60 days when stored in open while those stored in polyethylene bags at 8°C did not germinate at all.

The moist storage appears to be successful for preservation of recalcitrant seeds. Here seeds are imbibed with either germination inhibitor or fungicides for preventing germination and pathogens during moist storage. Seed treatment with 1% 8-hydroxy quinoline sulphate and storage at 15°C could preserve the seed viability for one year (Doijode 1995).

Citrus

Citrus seeds are either mono- or polyembryonic in nature. They are largely Utilized for raising rootstocks as well as in breeding programme. Lime (Citrus aurantifolia) seeds are small, oval and polyembryonic with white cotyledons. Lemon (Citrus limon) seeds are ovoid and polyembryonic with 10-15% nucellar embryos and white cotyledons. Sweet orange (Citrus sinensis) seeds are obovoid, polyembryonic with white embryos. Mandarin (Citrus reticulata) seeds are small polyembryonic with green embryos. Pummelo (Citrus grandis) seeds are large ridged yellowish and monoembryonic. Grape fruit (Citrus parodisi) seeds are white, polyembryonic with white cotyledons.

Seeds of many Citrus species are viable for a short period under ambient conditions. The seeds belong to orthodox group and/or show intermediate storage behaviour and require special procedure for preservation. Seed storage in moisture proof containers at low temperature checks the deterioration by reducing metabolism. Seed storage with calcium chloride in polyethylene bags at 8°C resulted in 100% viability in Citrus karna and rusk citrange and 84% in Citrus jambhiri and Citrus limonia after 150 days (Krishna and Shankar 1974).

Acid lime seeds were successfully stored for 10 years at 5 and -20°C. Grape fruit seeds retained viability for 3 months when stored with charcoal at room temperature and for 4 months at 5-8°C (Chacko and Singh 1968). In Poncirus trifoliata and Citrus grandis, germination was the highest after 80 days in seeds stored in polyethylene bags at 4.4°C and 56-58% RH (Mallareddy and Sharma 1983). Seeds could be stored in fruit itself for short term preservation. Fruits of some lemon cultivars were successfully preserved for 40 days at ambient temperature and for 60 days at 5°C (Doijode 1984). Bajpai et al. (1963) also reported that seed storage in whole fruit of kagzi lime and sweet orange is better than in polyethylene bags.

Rambutan

Seeds are difficult to store under normal conditions (Ellis 1984) and at room temperature they lose their viability within a matter of days. They can be stored for a few weeks in moist saw dust or charcoal to which some aril juice has been added at 21-28°C (Chin 1975).

Jackfruit

Seeds are large oblong about 3 x 2 cm in size having thick gelatinous yellow covering and belong to recalcitrant group. Seeds remained viable for 15 days and gave 40% germination after 30 days of storage (Anon. 1966-67).

Litchi

Litchi seeds are dark brown, covered with white fleshy, juicy and translucent aril. Seeds are recalcitrant and should not be allowed to dry. When sown immediately after extraction, their germination was 100% which reduced just after one day storage in water and it was 75% on the seventh day and 30% on 15th day after storage (Ray and Sharma 1985). Seeds stored in moisture proof containers sprouted within 15 days and those stored for 10 days in polyethylene bags showed 50.7% germination (Hay and Sharma 1987). However, seeds can be preserved in fruit for short periods. Seed viability was preserved for 24 days by storing fruits in sealed polyethylene bags after treatment with benomyl (0.05%) and 6% wax emulsion (Ray and Sharma 1987).

Mangosteen

Seeds are recalcitrant and lose their viability rather quickly, especially when the thin membrane around the seed is damaged or when seed is placed in ambient or cool atmosphere (Hanson 1984). Storage of seeds in charcoal or moss at room temperature retains the viability for a few months (Cox 1976).

Conclusions

Seeds of several tropical fruit crops show recalcitrant storage behaviour and they do not withstand drying or are able to keep alive at low temperatures. Thus, they are difficult to store for longer period (Ellis 1984; Hanson 1984). In this context, more research is needed for extending storage life. However, it is desirable to preserve the germplasm in field as fruit repository. To supplement the conservation process, wherever, as far as possible seeds could be conserved by use of available techniques like moist storage, etc., for certain period.

Germplasm of Mango

Mango is one of the most important fruit crops in India. It has been in cultivation for almost last five thousand years. Apart from India, mango is also an important crop in the Philippines, Indonesia, Thailand, Myanmar, Malaysia, Sri Lanka, U.A.E., South-East Africa, South Africa, U.S.A., Mexico, West Indies, Brazil and tropical Australia. All the cultivated Indian mangoes belong to the single species Mangifera indica L.

The geographical distribution, phytogenetic trend, pollen morphology, chromosome number, cytogenetical and breeding behaviour indicate that the highest concentration of species of Mangifera is found in Malayan Peninsula followed by Sunda Islands and the Eastern Peninsula comprising Myanmar, Thailand and Indo-China. Occurrence of wild species like M. sylvatica and M. caloneura in Myanmar, Assam and northeastern India and evidence of fossil leaf impression of

M. pentandra which looks similar to M. indica having edible fruits indicate that mango originated in this region. In India, more than thousand varieties of mango are available and these have arisen mainly as a result of open pollination.

It is to be admitted that most of the varieties lack in one or the other character. As opined by Naik et al. (1958), this very diversity has tended to limit commercial production of fruits of standard quality in India. In the recent past, efforts were made to exploit natural variability and breed new varieties for different purposes. So there has been a need to prevent the loss of varieties, and to conserve and use them in the breeding programme. Apart from intervarietal breeding programmes, progress has been made with regard to identifying the location of wild species of Mangifera and their utilization in breeding (Mukherjee 1985).

Mangifera Species—Present Status

It was thought earlier that the genus Mangifera comprises 41 species (Mukherjee 1949). Ding Hon (1978) opined that there are only 32 species. Tardieu (1962) added 4 species to this list from Indo-China. Ling (1983) added one species from China and in Malaysia two more were added by Kochummen (1983). However, Mukherjee (1985) reorganized the classification and recognized 39 species in the genus Mangifera. Subsequent surveys carried out by various workers in various parts have shown the existence of different races.

Yadav (1996) reported the occurrence of races of wild species of Mangifera in the northeastern region of India. Many of the seedling types, growing in the forest area of northeast region still have many primitive characters such as polyembryony and dwarf growth habit. Studies of wild mango trees in Manipur showed that primitive characters still persist and their mixing with the present day commercial varieties has not taken place (Yadav and Rajan 1993).

Some wild forms recorded in Manipur are similar to M. indica, but closer look at the flower and leaf characters showed that flower staminodes are absent and the disc is thick and tuberculate. This wild type resembles more with

M. caloneura (a species occurring in Myanmar) in having smooth ovary and stone not exactly kidney shaped. This also resembles M. siamensis because of the absence of staminodes. Another wild type with small fruits and big leaves was observed on Silchar-Imphal road which resembles the specimen reported by Sharma and Sen Chaudhary (1976) from Tripura, except that the base of the petiole in this type was swollen.

Mukherjee (1948) has reported five species, M. indica, M. andamanica, M. camptosperma (treated under M. gedebe), M. khasiana (now treated under M. sylvatica) and M. sylvatica to be indigenous to India. Another species M. nicobarica is said to occur in Nicobar Island. Areas of concentrated population of the trees of species have been identified for M. andamanica, M. camptosperma and M. sylvatica. M. indica is found all over the country as all the varieties in existence in the country belong to this species.

Distribution of Mangifera species

Country/Island	*Number of species*
India	5
Sri Lanka	2
Andaman Islands	3
Myanmar	6
Thailand	9
Indo-china	10
Peninsular Malaysia	19
China	1
Sumatra	11
Java	9
Kalimantan, Sabah, Sarawak	10
Bali	2
Philippines	8
Celebes	4
Moluccas	5
Timor	2
Irian Jaya, Papua New Guinea	2

Several of the Mangifera species face extinction. Mukherjee (1985) has listed the species under threat as follows:

1.	Endangered:	M. cochinchinensis
	M. flava	
	M. lagenifera	
	M. pentandra	
	M. superba	
2.	Vulnerable:	M. caloneura
	M. monandra	
	M. timorensis	
	M. zeylanica	
3.	Rare:	M. andamanica
	M. camptosperma	
	M. gedebe	

Although, there is awareness among the workers with regard to the utility of the species, hardly any effort has been made to characterize and to generate basic information regarding their crossability with the cultivated species. Screening of the species needs to be urgently carried out for pest and disease resistance. However, one of the main problems encountered in the conservation of species has been its precise climatic requirement. It has been our experience that whenever a species was introduced from its original place of growing, it has failed to acclimatize in the place of introduction. Hence, conservation of species needs to be carried out at the place of origin or in the place that resembles its place of origin.

Varietal Diversity

Apart from the species, there are about thousand varieties belonging to Mangifera indica (Mukherjee 1949). These varieties belong to two distinct classes *i.e.* polyembryonic types and monoembryonic types. About twenty polyembryonic types have been reported in India (Yadav and Rajan 1993) which has been the main centre of development of M. indica and its varieties. In India, there are three main centres of development of mango *i.e.* Lucknow, Saharanpur belt of U.P., Murshidabad area of

West Bengal and Hyderabad area of Andhra Pradesh (Yadav and Rajan 1993). The important commercial varieties.

Cultivated varieties of Indian mango have a long history. A large variation exists on the varietal front and varieties have different sizes, shapes, flavour and taste. Taking several characters into account like colour of emerging leaves, colour of panicle axis and laterals, size of flowers, intensity of pubescence of panicle and branches, 72 varieties from West Bengal were classified by Mukherjee (1948). Naik and Gangolly (1950) also described 135 south Indian varieties, taking fruit character as the main basis. Singh and Singh (1956) described 156 varieties of Uttar Pradesh based on primary, secondary and tertiary characters. Several workers described likewise the varieties collected as the germplasm in various centres (Yadav and Rajan 1993).

The north Indian varieties are generally alternate bearers and the south Indian varieties are regular bearers. However, there are some varieties which flower twice or thrice in a year or may continuously flower throughout the year, viz., Amino-Do-Phasla, Teenphasla, Baramasi types, Chiratpudi Goa (Royal Special). In Kanyakumari also continuous flowering is noticed.

Popular Varieties of Mango in Different Regions

Region	*Varieties*
Eastern	Himsagar, Fazli, Langra, Bombay
Western	Alphonso, Pairi, Rajapuri, Kesar
Northern	Dashehari, Langra, Bombay Green, Chausa
Southern	Neelum, Banganapalli, Totapuri, Mulgoa, Raspuri, Swarnarekha, Rumani, Badami

Although growth in mango is genetically controlled, the environmental interaction bring about the change in growth pattern under different agroclimatic conditions. Commercial varieties of different regions behave differently when grown in other agroclimatic zones. The variety Amrapali which remains dwarf under north Indian conditions does not remain so under south Indian conditions. Dashehari, Langra and Chausa rarely flower under south Indian conditions. Yadav and Singh (1985) opined that the north and south Indian varieties belong to two different groups representing two different ecotypes of M. indica, based on physiology of flowering.

Collection and Conservation

Mango in India has a long history of domestication and cultivation. The rich varietal wealth, however, has not been collected at one place. Hence, there is erosion of the genepool owing to deforestation, urbanisation and industrialization. Systematic attempts to conserve the genepool have been made in the past few decades and more than 1266 varieties were collected in various field genebanks (Chadha 1976).

In-situ Conservation

The species of Mangifera occur mainly as a complex biotic community in tropical humid forests, subtropical rain forests and dry forests/woodlands of Indo-Malayan biogeographic realm (Mukherjee 1985). For in situ conservation, the hilly regions of east Orissa, and forests bordering Myanmar in Manipur valley have been identified (Rajan et al. 1989; Yadav 1996).

Ex-situ Conservation

Most of the important commercial varieties grown in India as stated earlier have arisen due to natural cross pollination and selection. Germplasm collections have been made in the past. Although, substantial diversity of cultivated M. indica is in the field genebank, this does not contain wild species. Germplasm collections at many centres are duplicated and it has been noticed that varieties belonging to one region behave differently in other regions. Yadav and Singh (1985) opined that south and north Indian varieties belong to two different ecotypes of M. indica. Mustaffa (1993) confirmed the same by isozyme analysis. Hence, diverse collection has become a necessity to get actual performance. The major germplasm collection centres are listed.

Conservation in Vitro

One of the main constraints in the maintenance of mango in field genebank has been that it requires large area and funds. In this regard, in vitro conservation holds promise.

However, in mango, owing to the problem of regeneration of tissue, so far no success has been obtained. Iyer and Subramanyam (1989) have suggested pollen storage of important mango varieties. This will also help in speeding up the hybridization programme.

Utilization

Almost all the commercial cultivars belong to M. indica. However, a few commercial cultivars of South-East Asia belong to other species, such as M. altissima, M. caesia, M. foetida, M. griffithii, M. odorata, M. pentandra, M. sylvatica, M. zeylanica, M. laurina, M. lagenifera, M. cochinchinensis, etc.

For the improvement of mango, some of the wild species could be of great help, Iyer (1991) suggested that Mangifera species useful for crop improvement fall into two categories: (i) species having edible fruits with other desirable characters, and (ii) species that could act as gene donors for specific improvement like resistant to pests and diseases. If basic information like crossability become available then interspecific hybridization in mango can become a reality.

Major Ex-situ Collections of M. Indica Germplasm in India

Centres	*No. of accessions*
1. Eastern India	
Central Horticultural Experiment Station, Ranchi, Bihar	168
Agricultural Research Institute, Pusa, Bihar	148
Fruit Research Station, Krishanagar, West Bengal	59
2. Northern India	
Central Institute for Subtropical Horticulture (CISTH),	
Lucknow, Uttar Pradesh	565
Indian Agricultural Research Institute (IARI), New Delhi	
Horticulture Experiment and Training Centre, Chaubatia,	157
Ranikhet, Uttar Pradesh	254
Horticultural Experiment and Training Centre, Basti, Uttar Pradesh	

Contd...

Centres	*No. of accessions*
GB Pant University of Agriculture & Technology,	
Pant Nagar,	155
Uttar Pradesh	124
Fruit Research Station, Aurangabad, Maharashtra	130
3. Western India	
Regional Fruit Research Station (RFRS),	
Vengurla, Maharashtra	204
Agricultural Experiment Station, Paria,	
Gujarat Gujarat Agricultural University (GAU),	
Shahi Bagh,	151
Ahmedabad, Gujarat	109
MPKW, Rahuri, Maharashtra	86
MKV, Parbani, Maharashtra	80
4. Southern India	
Indian Institute of Horticultural Research (IIHR),	
Bangalore, Karnataka	174
University of Agricultural Sciences (UAS),	
Dharwad, Karnataka	54
Fruit Research Station (FRS), Sangareddy,	
Andhra Pradesh	632
Fruit Research Station (FRS), Kodur, Andhra Pradesh	148

Promising Characteristics of Some Wild Mangifera Species

Sl. No.	*Species*	*Special characteristics*	*Source*
1.	M. caesia	White pulp, sweet, fragrant	East Borneo
2.	M. decandra	Rootstock for waterlogged conditions	Sumatra, Borneo
3.	M. gedebe	Rootstock for waterlogged conditions	Borneo, East Katimantan, Sumatra, New Guinea

Contd...

Sl. No.	Species	Special characteristics	Source
4.	M. indica var. mekongensis	Fruits twice in a year and hence good parent in crossing	South Vietnam
5.	M. pajang	Fruit peeled like banana and could yield new varieties by selection	Sarawak, Borneo
6.	M. similis	Good parent for breeding freestone mangoes	Borneo, South Kalimantan

Hybridization

Hybridization is one of the methods to use the germplasm most effectively. Several hybrids and clonal selections have been released over the years with improvement in some characters. In any mango hybridization programme in India, following are the breeding objectives: (i) regular bearing; (ii) dwarf tree habit with precocity; (iii) attractive and good sized fruits (350-450 g) with good quality (tasty, firm flesh without fibres); (iv) resistance to major diseases like malformation, powdery mildew, anthracnose and insects like fruit fly and stone weevil; (v) freedom from jelly seed and internal break down; and (vi) good keeping quality.

Germplasm evaluation carried out over the years has pointed out several sources of desirable characters in the germplasm. Varietal evaluation carried out has shown that the varieties Lazzat Baksh, Neelum, Banganapalli, Kalapady, Kadiri and Padiri were found to be good source for regular bearing. Iyer and Subramanyam (1986) have reported that the variety Creeping can be a good source for dwarfness. The recently developed hybrids like Ratna and Arka Puneet can be used as a good source for freedom from spongy tissue. Naseeruddin Hyder and Laxminarayana (1979) grouped Navneetam and Lord as tolerant to hoppers. Hashmath Unnissa (1979) classified the varieties Azam-us-Samar, Chambeliwala, Chausa, Karnuli Mulgoa, Totapuri and Vanraj as resistant to powdery mildew, while Totapuri Red Small was found to be tolerant under Punjab conditions. Peethambaran and Aravindakshan (1979) reported that the variety Alphonso showed maximum tolerance to sooty mould at Trichur. Likewise varieties suitable for processing have also been identified. The variety Dashehari was found to be good for nectar making (Roy et al. 1972; Kalra et al. 1982) and Mallika was noticed to be good for canning of slices (Khundiya and Roy 1988). The variety Arka Puneet and Arka Anmol were found to be good for canned juice preparation and Saheb Pasand to be ideal for processing, having a TSS of more than 30 Brix under Lucknow conditions.

With the existence of large number of varieties, there is a good prospect for utilizing this varietal wealth in the breeding programme (Sharma 1987). However, mango being highly heterozygous, it is imperative that large number of progenies are to be raised to develop and isolate improved genotypes.

Future Thrust

Inspite of much work done, still the germplasm has not been fully utilized, the rich natural variability remains to be exploited both at the species and varietal level. Future research work needs to lay stress on the following:

1. Use of isozyme analysis for characterization so that duplicates could be identified.
2. Intensification of disease and pest screening to identify with varieties or species tolerant/resistant to various biotic and abiotic stresses.
3. Storage and exchange of pollen needs to be standardized, so that conservation as well as breeding can be speeded up.
4. Use of tissue culture for standardizing embryo culture and exploitation of protoplast fusion for somatic hybridization so that other Mangifera species can be used in breeding programmes where incompatibility barrier exists.
5. Use of polyembryonic varieties in the breeding programme by creating more variability in the polyembryonic types.

9

Management of Fungal Diseases in Tropical Fruits

In India, the diverse agro-climatic conditions favour cultivation of more than five kinds of fruits and nut crops, majority of which are grown in the tropical region. The major commercial tropical fruits are grapes, banana, mango, citrus, papaya, pineapple and sapota. Although the area under fruit crops has increased in the last three decades, but still it is only a little over one per cent of the total cultivated area.

The increase in area alone cannot maximise the yields unless and until per unit area production is increased. It is possible only through the development of high yielding varieties and the knowledge about the factors responsible for the crop losses. Among these factors, plant diseases specially fungal diseases, play a major role in reducing yields of horticultural crops in the tropics.

It is estimated that the production could be increased at least by 28% if the crop could be protected against various diseases. Many a time, most of the diseases lead to the complete crop failure if it is not checked in time. Many of these diseases are transmitted through the planting material. The phenology of fruit crops which are perennial is different from other field crops which are seasonal in nature and hence the pathogens keep on perpetuating on the plant itself in fruit crops.

Further, the knowledge about etiology and epidemiology of diseases is most important which would help in profitable cultivation of fruit crops and the measures adopted thus, in turn, will improve the economy. This knowledge is also important for optimal management of field genebanks of various fruit species. The available relevant literature on some of the most important crops has been screened and is presented here.

Fungal Diseases in Tropical Fruits

Mango

Mango (Mangifera indica L.) is an important fruit and is subjected to a number of diseases at all stages of its development *i.e.* from nursery to the consumption of fruits. All the plant parts, namely, trunk, branch, twig, leaf, petiole, flower and fruit are attacked by different pathogens. Some of the diseases are very severe on plant and produce alike and have become a limiting factor in profitable mango orcharding. Bloom blight caused a complete failure of the crop in Florida and Brazil, black spot in South Africa and powdery mildew and Black-tip in India, etc.

Many types of agents cause diseases in mango and of these fungi cause the largest number of diseases, while bacteria, algae, angiospermic parasites and nutritional deficiencies are the other causal agents of mango maladies. Virus, which causes severe infection in other fruit crops, is luckily not reported so far on mango.

Many of the mango diseases have been investigated and also measures to combat them have been worked out in different countries. However, our knowledge about them is still imperfect and inadequate. Vast areas in South-East Asia - the home of mango, still remain unexplored from disease point of view and may possibly offer many other problems for investigations. The information available on some of the important diseases is presented below.

Powdery Mildew (Oidium mangifera Bert.)

First reported from Brazil in 1914 by Berthet, later on it has been recorded from several countries of the world. In India, it has been reported from Uttar Pradesh, Punjab, Maharashtra, Gujarat, Madhya Pradesh, Jammu & Kashmir, Haryana, Andhra Pradesh, Karnataka and Tamil Nadu. The losses caused vary from 5-20 per cent depending upon the weather conditions.

Prakash and Srivastava (1987) reported 30-90% losses in Lucknow. This is a serious problem in hills, valleys and tarai regions of Uttar Pradesh. Serious outbreaks have been noticed in Uttar Pradesh in the last 6-7 years. It attacks the leaves, flower scales, buds of tender flower heads, axils, stalks and fruits. It manifests itself as wefts of white mycelium on the affected parts.

The whole surface is later on covered with a powdery substance which is blown away by even a slight disturbance caused by winds. The fungus is ectophytic, remaining mostly on the surface and draws its nutrition from the plant cells through the haustoria which penetrate the epidermal layer and reach the cells. The affected fruits do not grow in size and may drop before attaining pea size. In Bangalore area, the disease has been found to be prevalent almost throughout the year on varieties like Pairie and Alphonso.

In Indo-gangetic region the infection is occasionally found on malformed inflorescences and there is a likelihood of its survival in the sub-mountainous regions of Uttar Pradesh. Warm temperature with heavy morning dew and cloudy weather favour the disease development (Kulkarni 1924). The minimum, optimum and maximum temperature for conidia germination is stated to be 9, 22 and 30.5°C respectively (Uppal et al. 1941). The disease appearance and incidence is greatly influenced by the sunshine (hours) per day and it is inversely proportional to the disease having a correlation coefficient of r =0.5943 (Gupta 1985). The other factors influencing the disease was relative humidity at 1430 hours (Rawal and Saxena 1992).

The most effective method of control is by dusting the plants with fine sulphur (250-300 mesh) at the rate of 1-3 lb. of sulphur/tree (Cheema et al. 1954). The first application may be soon after flowering, second 15 days later followed by a third one. Spraying of Guesarol (405.50) has been found successful in Tamil Nadu, Andhra pradeeh and Maharashtra. Several fungicides, such as Cosan (0.2%), benlate (0.2%), wettable sulphur and DDT (2:1), karathane or morestan, carbendizim (0.1%), 0.2% microsul and butrimate, Baycor, Calixin, Anvil (0.05%), Systhane (0.05%), Saprol (0.2%) were found to be effective in controlling the disease (Sohi and Sridhar 1973; Lingraj 1969; Fundacion 1972; Prakash and Singh 1982; Rawal and Ullasa 1985; Rawal and Saxena 1989).

Anthracnose: This is caused by Colletotrichum gloeosporioides Penz. = Glomerella cingulata (Sten.) Spould and Shrenk., Gloeosporium mangiferae P. Henn and is the most important disease wherever mango is grown. The disease develops on all the tender parts of the plant and is especially serious on tender twigs, flowers, flower stems and fruits in storage.

In India, it is widely distributed in Bihar, Punjab, Maharashtra, Gujarat, Kerala, Uttar Pradesh, Rajasthan and Karnataka. The disease produces leaf spot, blossom blight, wither tip, twig blight and fruit rot symptoms. Young leaves when attacked wither and dry up. Sometimes only the edges of the leaves are attacked; their margins darken, dry up, may fall out, giving the leaf a rugged appearance. It produces a wither tip of young tender twigs. It also results in a die-back which appears as blackish spot on the growing tips. The affected branches ultimately dry up and infections keep on penetrating down.

On fruits, most of the infection takes place from the start of the blossoming period until the fruits are more than half grown. The spots appear near the stem end as small brown areas that enlarge rapidly and become black. In some cases, the areas involved are in the form of streaks running down from the stem end. The affected areas usually crack and sink slightly.

The decay is confined to the skin of the fruit except in late stages when it penetrates the flesh in shallow areas. The latent infection (during ripening) is carried from the field and develops further which causes rotting in the storage. Healthy fruits develop infection after coming in contact with diseased fruits. The inoculum remains on dried leaves, defoliated branches and mummified flowers and flower brackets. The severity and prevalence of the disease is influenced by excessive rains, heavy dews or high humidity during critical infection period (Doidge 1932). The temperature range for disease development

10-30°C, and the R.H. 95-97% are highly congenial. Spotting of leaves and twigs which become severe in damp weather are generally not serious enough to cause wither tip and die back in full grown trees (Kausar et al. 1960). The secondary spread is through rain drops.

Only one variety, namely, Edward has been reported to be resistant (Brooke and Olmo 1957). In Uttar Pradesh, anthracnose was effectively controlled by spraying with carbendazim (1%) or Topsin-M (0.1%) or chlorothalonil (0.2%) at 14 days intervals until harvest. Benlate (0.2%) and Dithane Z-78 (0.2%) are extremely toxic to fungus in culture.

However, these have not been tested in the field (Bose et al. 1953). Spraying of Micop, Blitox and dithiocarbomate were found effective (Lingaraj 1969). Bordeaux mixture (1.0%) has been recommended (Gadre 1979). Application of Captan (3 g/l) or Phaltan (2g/l) are also recommended (Oriux 1961).

For the control in storage, the management strategies recommended to control anthracnose include cultural practices, tree management, varietal selection and protective sprays using curative fungicides with advantage that sprays need only to be applied after infection occur and using prochloraz and disease detection instrument known as mango anthracnose estimator (MAE) (Fitzell and Peak 1985). Hot water treatment has also been recommended by various workers which is 51.5°C for 15 minutes (Pennock and Maldonaldo 1961); 50.5-55.5 °C for 5 minutes (Soot and Segall 1963); 54-55°C for 15 minutes (Tandon and Singh 1968) and 55°C for 5 minutes added with Benomyl or TBZ (Spalding and Reeder 1972).

Anthracnose would be controlled by dipping the fruits in various fungicides or exposing them to ammonia, sulphur dioxide and carbon dioxide gases and also by hot air treatment (Tandon and Singh 1968). The fruit dipping in benomyl solution (9500 ppm a.i) or thiobendazole (1000 ppm) immediately before storage reduces the disease to 5 per cent from 29 per cent (Sohi et al. 1973). Pre-harvest spraying with Bavistin (0.1%) or Topsin-M(0.1%) or Prochloraz (0.1%) followed by dipping in the same after harvesting have also been found to be effective.

Mango Malformation (Fusarium moniliforme var. subglutinans Sheldon)

It is the most important malady of mango and was first reported by Burn (1910). It causes the losses from 50-80 per cent. It is also reported to be due to other disorders like physiological and acarological. Indications of its virus nature has also been postulated (Vasudeva 1959). The maximum incidence of the disease has been recorded in the north, northwest and north-east parts of the country.

The incidence is, however, less in western and southern India. There are two types of symptoms, namely, floral malformation and witches broom or bunchy top or vegetative malformation with a proliferation of infected tissue. The flowering panicles turn into a compact mass of flowers. This compact mass is very hard and not soft like normal panicle. Individual flower is greatly enlarged and has a large disc. The percentage of bisexual flowers in malformed panicles is very low. In bunchy top, compact leaves are formed in a bunch at the apex of shoot or in the leaf axil. A similar bunch consisting of small rudiments crowded together on short shootlets is seen in vegetative malformation in which the growth of shootlet is arrested. Vegetative malformation is more pronounced in young seedling and seedling trees. The malformed heads dry up in black masses and persist on the trees for a long time. The malformed inflorescences contain more of endogenous cytokinins than healthy ones (Nicholson and van Staden 1988).

The variety Zebda has been reported to be resistant in Egypt (El Ghandour et al. 1979). Considerable incidence reduction has been reported by spraying 100-200 ppm NAA during October (Majumdar 1977). Use of disease free planting material and prophylactic spray of insecticides and fungicides can keep the orchards healthy. In areas with less than 5-10 per cent infection, pruning of diseased plants should be made compulsory; pruning of diseased parts along the basal 15-20 cm apparently healthy portions. This is followed by the spraying of Bavistin (0.1%) or Captaf (0.2%). Partial control of malformation has been achieved by spraying mangiferin-Zn_2 and mangiferin-Cu2+chelates (Chakrabarti and Ghosal 1988).

Stem End Rot (Botryodiplodia theobromae Pat.)

It is an important disease of ripe mango. This fungus also causes die-back symptoms in orchards. It is known to occur in India, Myanmar, Sri Lanka, the Philippines, Mauritius and the U.S.A. In India, it is recorded from all the mango growing areas. The disease is characterized as the dark epicarp around the base of the pedicel in the initial stage.

In the next few hours, the affected area enlarges to form a circular, black patch which in humid atmosphere extends rapidly and turns the whole fruit completely black within two or three days. The pulp of the diseased fruits become brown and somewhat softer. The diseased fruits showed increased phosphorus and reduced ascorbic acid contents (Desai and Pathak 1970). Pectolytic enzyme activity was also lowered in such fruits (Pathak and Srivastava 1969). The disease is carried from the dead twigs and bark of the trees harbouring the pathogen and with the onset of rains the orchards get contaminated.

Dipping of mangoes in 6 per cent borax solution at 43°C for 3 minutes reduces the incidence (Thomas and Dalai 1968). Fruit dipping in Aureofungin and wrapping was also found effective (Dharamvir et al. 1967). Pathak (1980) suggested to harvest mangoes on clear dry day. Hot water treatment at 53°C for 10 minutes (Quimio and Quimio 1974) and at 51 to 55°C for 15 minutes (Liu 1985) have been recommended elsewhere. Care should be taken to prevent snapping-off of the pedicels. Injury should be avoided to fruits at all stages of handling. Spalding (1982) in his experiments found thiobendazole and thiophenate methyl to be the most effective ones. Recently, Rawal and Ullasa (1988) have reported that Bavistin (0.1%) or Topsin-M (0.1%) or chlorothalonil (0.2%) spraying in the field before harvesting give very effective control.

Die-back (Botryodiplodia thiobromae Pat)

It is a descructive disease of mango and is known to occur in India and other mango growing countries. The disease is manifested by discolouration and darkening of the bark some distance from the tip of the twigs. The dark area advances and young green leaves start withering first at the base and then extending outwards along the vein. The affected leaves turn brown and the margins roll upward. At this stage, the twig or branch dies, leaves shrivel and fall, and this may be accompanied by exudation of gum.

The infected twigs show internal discolouration. Brown streaks of vascular tissues is seen on splitting the twigs lengthwise along the long axis. In early stages, epidermal and sub-epidermal cells of twigs appear slightly shrivelled. The areas of cambium and phloem show brown discolouration and yellow gum like exudate flows out of the cells.

The disease is known to be enhanced by a beetle, Xyleborus affinis (Batista 1947). Relative humidity above 80% and temperature of 25 to 31.5°C, and rains enhance disease development. The die-back is also caused due to Colletotrichum gloeosporioides. Both these pathogens also attack fruits which get spoiled during ripening.

For effective control of the disease, pruning and destruction of infected twigs is the foremost practice. Spraying the trees periodically with copper oxychloride sulphate is recommended by Alvarez and Lopez (1971). Pasting of trees with a mixture of oil and 5% phenol was found effective (Batista 1947). Sprays of carbendazim (0.1%) or methyl thiophenate (0.1%) or chlorothalonil (0.2%) at fortnightly interval during rainy season is important. Sometimes short hole borers also predispose the trees to infection and hence proper insecticides are also to be sprayed. Healthy twigs should be selected for grafting of seedling during propagation.

Leaf Blight (Macrophomina mangiferae Hingorani & Sharma)

It was first reported from Brazil to be caused by Macrophomina sp. by Lacca (1922). In India, it was reported by Patel et al. (1949). Later, it has been reported by various workers from Delhi, U.P. and Gujarat. The other pathogens causing the disease are grey blight (Pestalotiopsis mangiferae), phoma blight (Phoma glomerate), and twig blight and fruit rot (Phoma sp.).

The disease usually appears as yellowish pin head like spots on leaves and twigs of the affected plants. These gradually enlarge discolouring the surrounding tissues which first become brown and then dark brown with slightly raised and brown purplish margins and later ash coloured due to the appearance of pycnidia.

Spots are round to start with but later become oval or irregular in size depending upon environmental conditions such as humidity and temperature. Infection is mainly observed on leaves and rarely on stems. On stem, the lesions are elliptic which later girdle the stem at the point of infection. On fruits, water soaked circular lesion are produced which enlarge rapidly and cause rotting.

The pathogen can survive for more than a year on the leaves of mango (Hingorani et al. 1960). The varieties, namely, Khandeshi, Borasio and Asal Damido are reported to be resistant (Desai and Patel 1963). Field sanitation is by collecting and destruction of diseased parts. Spraying of Burgandy mixture, Perenox, lime sulphur and dithane have been recommended by Hingorani et al. (1960).

Red Rust (Cephaleuros mycoides Karst)

This or red spot is a common algal disease on the mango in the tarai and in the other humid regions of India. Its occurrence has been reported from Bihar, Karnataka and Uttar Pradesh. The alga attacks the foliage, bark and twigs of the host plant and is both parasitic and epiphytic in nature. In serious infections, the bark becomes thickened, the twigs get enlarged and remain stunted and the foliage becomes sparse and finally dries up.

Initially the spots are greenish-grey in colour and velvety in texture, but later on the surface bears reddish-brown appearance. Algal spot is circular to irregular in shape, slightly elevated and with usually 2 mm in diameter, though in some cases it may be as much as 1 cm. Bordeaux mixture (6:6:100), Supramar, Fytolan, Blitox-50 or lime sulphur are recommended (Pathak 1980).

Black Spot (Xanthomonas compestris pv mangiferae indicae)

It is one of the serious diseases and is reported from many countries. In India, it is reported from Maharashtra, Delhi, Uttar Pradesh, Tamil Nadu and Karnataka. It affects leaves, petioles, fruits and tender stems. Numerous small angular water soaked lesions appear in groups towards the tip of the leaf blade.

These are first light yellowish but later turn dark brown to black and get surrounded by a distinct halo. Several lesions are often found to coalesce forming large necrotic spots. On young fruits, water-soaked lesions develop which also turn dark brown to black. Infected fruits may show skin cracking and badly affected ones dry prematurely. Infection by wounding produces diseased spots on tender stems, petioles and stalk bearing fruits. These spots turn into deep longitudinal scars accompanied by appreciable amount of bacterial gum-like ooze. The incubation period on leaves varies from 2-14 days (Patel et al. 1949).

In India, the disease remains dormant during November to March due to low temperature (11.8-22°C) and the leaf infection is considerably reduced by the fall of the infected leaves. Kent mangoes showed close relationship of rainy season with incidence of bacterial leaf spot. One day's rainfall did not affect the disease appearance, whereas four days of rainfall led to the 36.9% disease incidence.

Twig canker persists and initiates fruit infection. The disease spread is rapid during the rains and becomes severe in July-August. Five application of Bordeaux mixture (4:4:50) with spreader was recommended by Wager (1937), whereas four sprays of 6:6:50 Bordeaux mixture was found to be effective by Marloth (1947). Sprays with fungicides such as copper containing materials and agrimycin (Viljoen and Kotze 1972) have also been advocated. Shekhawat and Patel (1975) recommended orchard sanitation by way of removal of infected materials and seedling treatment as preventive measures. Agrimycin-100 proved best in arresting the disease development. Plantomycin (200 ppm) followed by agrimycin + bavistin (1000 ppm) was reported to be the best against this disease. Streptomycin sulphte followed by aureofungin have been recommended by Prakash and Raooff (1985) to control bacterial disease of mango.

Black Mould Rot (Aspergillus niger Van Teigh)

The affected fruits show yellowing of base and development of irregular, hazy, greyish spots and coalesce into dark brown or black lesions. The mesocarp of the rotted areas becomes depressed and soft. The diseased fruits show rapid decrease in ascorbic acid content (Ghose et al. 1966). The pathogen infects injured fruits (Srivastava 1968). Totapuri variety is highly susceptible (Patel 1972). A fruit dip treatment with benlate at 1500 ppm can control the rot (Bhargava and Singh 1975).

Soft Rot (Rhizopus arrhizus Fisc.)

This pathogen alone is reported to be responsible for the decay of 6.3 per cent fruits in Delhi markets (Thakur and Chenulu 1966). The pathogen incites a typical soft rot. Development of disease is slow and less intensive at low temperature (7-8°C). The disease develops rapidly between 20 and 40°C (Thakur 1972). Thakur and Chenulu (1970) found that 2-aminothiozole with 2-aminopyridine each at 7 per cent concentration protected mango fruits for 20 days.

Some of the other post-harvest diseases and pathogens are Alternaria rot (Alternaria tenuissima (Neer. Ex. fr.) Wiltshire), Sclerotium rot (Sclerotium rolfsii Sacc.), dry rot (Boothiella tetraspora Lodhi and Mirza), brown rot (Pestalotiopsis mangiferae P. Henn. and P. glandiola (Cast) Stey), fruit rot (Phomopsis mangiferae Ahmed) and black spot, Actinodichium jenkinsii (Uppal et al. 1941).

Black Tip : This disease is peculiar to India and does not occur in any other country. It has been reported to occur in regions of West Bengal, Uttar Pradesh, Bihar and Punjab in orchards located in close proximity of a brick kiln. South India is free from this disease. The symptoms manifest as small etiolated area at the distal end of the fruit, which gradually spreads, turns nearly black and covers the tip completely. Before the etiolation is complete, isolated greyish spots appear which become dark brown, enlarge and coalesce into a continuous necrotic area. The brown discolouration spreads to the neighbouring parenchyma and the deposits also appear in the ducts. The browning and deposits gradually spread throughout the mesocarp and the affected cells distintegrate and coalesce into a dead tissue. In severe cases, the necrosis extends to the endocarp.

The practical control measure lies in keeping the brick kilns away from the mango orchards. The use of telescoping chimney 12-15 m high has also been recommended (Sen 1943). Spraying of borax (0.6%) at 10-14 day intervals has been recommended (Reddy and Kapoor 1965).

Soft Nose : The problem was first observed as physiological breakdown in cvs. Kent and Haden by orchardists in 1950 in Florida. The soft nose is characterized by hallowing of the green skin in the areas between the apex and the stigmal point. A lack of firmness in diseased fruits can also be felt with experience. Upon cutting the flesh on the ventral side towards the apex of some soft nose fruits appear to be over ripe, while on shoulder and dorsal side it is unripe (Young 1957). The over ripe flesh surrounds a mass of yellowish to brown tissue which is firmer than the surrounding affected tissue and is bitter in taste.

Young (1957) concluded that the disorder is not due to the microorganisms. However, Young et al. (1963) reported the loss of only few percent in fruits harvested from trees grown on calcareous rock soil, but on acidic soil, sandy soil it is common for 15-20 per cent incidence. According to Young et al. (1963), soft nose disorder increased with N fertilization but was reduced to greater extent by increasing Ca level of the soil. Nitrogen fertilizer in sandy acidic soil increased the incidence whereas in calcareous soil there was no effect.

To overcome this disorder, the high N level of soil should not be allowed in the sandy or acidic soil and proper Ca level should also be maintained in soil. Early picking is also suggested (Young and Miner 1961).

The other disorders are little leaf caused by zinc deficiency, tip burn due to moisture stress and salt accumulation in soil and leaf scorch due to accumulation of chloride ion in leaf.

Phanerogamic Parasites (Dendrophthoe (Loranthus) viscum) : The mango trees are severely affected by this parasite particularly in neglected orchards. They are commonly present on the trunk or branches of the trees. The foliage of the infected tree is sparse, reduced in size and its bearing capacity and quality of fruits is considerably reduced. The fruits of this parasite are baccate with viscous mesocarp and attract the birds by its attractive colour. The seeds of this parasite are spread through bird and animal droppings. Such seeds, germinate and produce haustoria. Sometime tumour like swellings are observed on the affected tissue. There are many other species of parasites which also attack mango trees. All epiphytic plants can be removed easily in early stages before their aerial roots penetrate the ground and surround the host trunk.

Some more diseases reported by various workers attacking the mango crop in the field are leaf spots due to Rhizoctonia bataticola (Taub) Butt., Cercospora mangiferae indica Munjal Lal & Chona, Alternaria alternata Fr. & Keissler, Phoma sorghicola (Sacc.) Boerema. Dorem and Venkast; Scab (Elsinoe mangiferae Bit & Jen; black band (Rhinocladium corticolum Massee); pink disease (Pellicularia solmonicolor (Berk & Br.) Dastur); root rot and damping-off (Rhizoctonia solani Kuhn) and wilt due to Verticillium albo-atrum Reinke and Berth and Fusarium solani (Mart.) Sacc. There are several more fungi causing different diseases but the above are the most important among them.

Conclusions

Mango is attacked by various types of maladies (Rawal 1990). The epidemiology of different diseases is not fully understood and hence it should be taken up on priority basis to correlate various factors playing a direct or indirect role in disease epidemics so that the model for disease prediction can be worked which will be helpful in disease forecasting. Etiology of many problems is not yet known, thus its understanding is of utmost importance. Losses caused by individual problem or a combination of problems is not reported anywhere and it has to be determined. Some of the control measure given in the literature are very old and in the process some chemicals are either not available or they are outdated and since new fungicides/chemicals are being developed, their testing should be a continuous process to find out the best chemical for the control. This is specially important in maintaining clean and healthy field genebanks.

Citrus (Citrus spp.)

Citrus ranks the third place among the fruits grown in India and thus assumes considerable importance. It grows much better in tropical and subtropical conditions and there is an increase in area under citrus cultivation from 90.7 thousand hectares in 1961 to 279.4 thousand hectares in 1988. However, it is very sensitive to climatic changes and the varieties under cultivation vary from region to region. Sweet oranges and mandarins constitute the bulk of area under citrus fruits. Out of these two, mandarins occupy the most prominent place.

Due to their large scale cultivation, several serious problems have also cropped up. Among them the disease problems caused by many fungal, bacterial, viral, viroids and mycoplasma like organisms are important as they cause a huge economic losses and decline of the trees. Some of the fungal diseases like foot rot, root rot, gummosis, leaf fall and fruit rot, twig blight, powdery mildew and scab, etc., are the alarming diseases of citrus. In view of this, the present review is being compiled to gather more information available in the literature.

Die-back or decline is the most important problem at present being faced by the country (Aiyappa and Srivastava 1967). Decline is a symptom and not a disease. Earliest symptoms are dulling of the foliage, wilting of the leaves or delayed flushing of the tree. These symptoms are usually followed by chlorosis of leaves resembling zinc deficiency, but with a difference of showing speckling with green dots of the size of pinheads to nailheads.

Trees with pinhead speckles decline faster than the trees without them. Successive flushes produced on affected branches are erect, leathery, strap-shaped and chlorotic. Small percentage of mature fruits are reduced to the size of golf balls and contain columellas and aborted seeds. As defoliation progresses twigs begin to die-back. Trees do not die completely but sooner or later are removed because they become unproductive. Many factors, such as fungi, viruses, bacteria, nematodes, insect pests, cultural malpractices, unsuitable soil pH and high water table, etc., are responsible singly or in combination for the decline of citrus in India.

Among the various agencies that cause decline (Chadha et al. 1970), the diseases caused by fungi, viruses and bacteria are more predominant. Though the malady is in existence since 18th century, it has assumed serious proportions in the last few decades. Among all the Citrus species, varieties in commercial cultivation are the most susceptible. In Florida (U.S.A.), the disease is preceded by wilt and is caused by Fusarium oxysporum Schlecht. (Timmer et al. 1979). Apart from this, the following fungal diseases are important in different parts of the country.

Gummosis, Leaf fall, Root rot, Foot rot or Fruit rot (Phytophthora spp.)

This occurs in all the citrus growing areas especially in the high rainfall areas. Even then no comprehensive survey of the disease has been made so far. There are some reports on the prevalence of this disease, viz., in South India by Rajan and Aiyappa (1944), Ramakrishnan (1954) in Maharashtra, Gujarat; and Mysore by Cheema and Bhat (1928), Frazer and Singh (1966), Kapoor and Bakshi (1967) in Punjab; and in Assam by Chowdhuri (1951).

It is caused by more than one species of Phytophthora, *i.e.*, P. nicotianae var. parasitica (Dastur) Comb, is widespread in Assam and is often associated with Fusarium lateritium Nees (Chowdhuri 1951). P. nicotianae var. parasitica is also predominant in Karnataka, whereas P. palmivora (Butl) is prevalent throughout. Phytophthora produces symptoms of decline through: (1) a rotting of the rootlets, (2) a girdling of the trunk and a dropping of blighted leaves. The plants usually blossom heavily and die before fruits mature. The fruits lying on the ground are liable to invasions by the pathogens and develop brown rot. P. citrophthora has also been reported to cause the foot and root rot.

Symptoms : Certain parts of trees are more susceptible to Phytophthora than others and the different expressions of the disease syndrome vary accordingly. The disease manifests itself in the form of large water soaked patches on the basal portion of the stem near the ground level or it may spread further. Bark on the lower part of trunk is more susceptible to infection than the bark below the ground level.

In some very wet areas, the infection is observed quite high on the trunk and even on limbs. The gumming is sometime noticed on higher branches also. Initially, the bark is firm but with ageing of the lesion it shrinks and cracks, leading to shredding in lengthwise strips as it dries. The necrotic area is at first flush with the healthy bark, but with time it shrinks and becomes sharply delimited. The live bark at the advancing margins frequently develops a callus roll that checks further spread particularly in an upward direction. Symptoms on leaves are similar to those produced by Phytophthora infestans on the foliage of potato. Blighted leaves drop, and if defoliation is heavy the impression is given that the tree is in decline. Penetration of trunk tissue may be by way of wounds or by way of growth on cracks that develop with the emergence of adventitious shoots near the bud union.

The various names given to the different types of damages caused by Phytophthora are as follows; when the infection is continued to the feeder roots it is called feeder or fibrous root rot. If the damage to the root system is more extensive and yet does not extend upwards to the crown roots it is called root rot. The damage caused to the bark of crown roots and or bark of the trunk is called Phytophthora gummosis collar rot or foot rot.

Epidemiology : Leaves and fruits may also be attacked by Phytophthora nicotianae var. parasitica during prolonged periods of rainy weather and high humidity. The attack is more serious on mandarin. Oat meal-agar with yeast or beef extract gave maximum vegetative growth, whereas for sporangial production rain water was the best (Rao 1984). Collar rot becomes more severe in black soil than in light soils (Rao 1954). P. palmivora thrives best at 25-28°C. A soil pH of 5.4-7.5 favours the disease (Aiyappa and Srivastava 1967). Mosambi and Pummelo are more susceptible than mandarin. Lime and Jambheri are resistant. According to Raychaudhuri and Lele (1970), Troyer Citrange and trifoliate oranges seem to offer protection against brown rot and gummosis incited by Phytophthora species. Invasion of host tissue is by motile, free swimming spores called zoospores. Water (not just humidity) is required for sporangial formation, dispersal of zoospores as well as for spore germination (Whiteside 1973). Brown rot was greatly influenced by the rainfall and not by spacing (Timmer and Fucik 1975).

Role of Root Stock : The predominance of resistant and moderately resistant clones was observed in pummelo, sweet lime, trifoliate orange and trifoliate orange hybrids (Grimm and Hutchinson 1977). In their earlier studies conducted in 1973, it is reported that Citrus macrophylla, C. miaray, Severinia buxifolia, Cuban shaddock, Ogami Pummelo and Sour orange were rated resistant. In Turkey, Carrizo Citrange has been reported to be highly resistant (Cinar and Tuzcu 1976).

Control Measures : Adoption of tolerant rootstocks like sour orange, Cleopatra mandarin, trifoliate orange, Rangpur lime, etc., had been advocated by Chadha et al. (1970). A variety resistant in one region may be susceptible in other. The use of nursery stock grown under Phytophthora free conditions and prevention of water standing in contact to the susceptible portions of the bark above the bud union by way of good drainage. The disease can further be prevented to some extent by dusting of walls and bottom of planting pits with zinc-copper lime in ratio of 5:1:4 (Ramakrishnan 1954).

The exposed wood can also be covered with zinc copper lime (3:1:3.5) wash or bordeaux paste which should be followed by application of white lead or any wound dressing agents (Ramakrishnan 1954). Desai et al (1966) achieved complete control of spraying Khagzi lime trees infected with P. citrophthora with aureofungin 3 g/30 ml of liquid soap/30 gallons of water applied 2-3 times. Somani and Patel (1970) got good control of gummosis incited by P. nicotianae var. nicotianae by aureofungin as soil drench and spraying of foliage twice at an interval of one month. Drenching of Ridomil MZ (0.2%) has also been found effective against gummosis. Captafol (3 g/l), copper ammonium carbonate and cupric hydroxide are effective for foot rot (Timmer 1977). Dipping of seedlings in captan solution before planting is recommended in U.S.A. (Grimm and Whidden 1966). Fumigation of nursery bed with methyl bromide has proved most effective to avoid nursery infection (Grimm and Alexander 1971). Leaf fall and fruit rot could be controlled by spraying bordeaux mixture (1%) or bordeaux mixture plus tin sulphate or difolatan (0.3%) (AICFIP 1982). In addition, spraying of Ridomil MZ and Aliette (0.2%) have also been reported effective (Rawal 1990). Diseased leaves and fruits may be collected and burnt or burried deeply.

Anthracnose : The disease is a serious problem and a limiting factor in U.S.A. and Carribean countries. In India, the disease has been reported from Punjab, Madhya Pradesh, Uttar Pradesh, Assam, Rajasthan and South India. The disease is also known as withertip which in recent years is responsible for decline in production of lime in North India, particularly in Kanpur area. Withertip, leaf-spot and fruit rot, all are caused by Colletotrichum gloeosporioides Penz. (Pathak 1980).

Symptoms : Anthracnose attacks the young leaves, shoots, blossoms and fruits of small acid limes. Young foliage and blossoms are blighted and distinct lesions formed on leaves and fruits. Affected fruits frequently drop prematurely. Withertip is characterized by shedding of leaves and die-back of twigs. Leaves show light green spots which turn brown. On dead twigs, black dot like acervuli appear in concentric rings. The stem end of immature fruits results in fruit drop. In severe cases, branches show die-back and the tree dies in a few years. Symptoms of anthracnose appear on leaves, young shoots and tender fruits. On leaves, the necrotic spots show acervuli arranged in concentric rings. Dead parts of the twigs assume silvery grey appearance. Twigs show a slight gumming and a sharp line of separation between healthy and dead tissues. Affected buds fail to develop and affected fruits drop off. Often, the infected fruits develop reddish brown stain on the rind. The fungus has also been held responsible for russeting and tearstaining of rind. This type of symptoms can also be caused by rust mites. The disease is also caused by Gloeosporium limetticolum Clousen or G. foliicolum Nishida. Malta oranges and grape fruits suffer much more damage. C. gloeosporioides remains in a dormant condition in the dead twigs and branches. This leads to the twigblight. The infected hyphae produced by appresoria remain latent even after the fruit mature and produce anthracnose in oranges and grape fruit, if the peal is injured or fruits are over matured.

Control Measures : The disease can be minimised through proper management of orchards by proper irrigation, manuring, cultural practices and plant protection measures. Dead twigs should be pruned and destroyed, cut ends should be protected by bordeaux paste. Such trees should be sprayed with bavistin (0.1%) or captafol (0.2%) 3 times after pruning (AICFIP 1987). Chowdhuri (1936) advocated the pruning of diseased twigs in winter followed by spraying with bordeaux mixture in February and March, and September.

Singh and Sinha (1954) reported that spraying the fruits with careful deposition of spray at the stem-end, using 4:4:50 bordeaux mixture, in 3:5 lime-sulphur, 0.1% copper sulphate solution and 2 per cent formalin, considerably reduced fruit drop in preliminary trials. Satisfactory control of anthracnose of lime has been obtained through spray of trees by bordeaux mixture (5:5:50) or 0.33 per cent perenox. Spraying of bavistin (0.1%) or Topsin-M(0.1%) at fortnightly intervals are also effective during fruiting period.

Powdery Mildew (Acrosporium tingitanium Carter Subr.)

The disease is common in India and is serious on mandarin and sweet oranges in the sub-mountain tracts of Coorg, Nilgiris, Pulney, Wynad and Shevaroy hills in South India and in parts of West Bengal, Assam and Sikkim in the northeastern India. The disease has been reported from Bangalore on lemons in addition to above group (Rawal and Ullasa 1988).

Symptoms : All the aerial parts of the plant are attacked.

The disease is characterized by a whitish, mealy powdery growth on young leaves and twigs, which covers the entire leaf and young growing shoots. Even the small fruits also get infected. Such fruits shrink and become mummified. Affected leaves become distorted. The powdery coverage of foliage and twigs leads to leaf drop and die-back (Devarajan 1943). The affected fruits drop off prematurely and the yield is reduced.

In Coorg, the disease appears in a severe form between October and March. Damp mornings with a few hours of sunshine favour onset of the disease. The pathogen enters the epidermis and absorbs nutrients through haustoria. The mycelium covers the host surface. Conidia are disseminated by wind. Such conidia germinate ectophytic and later on become endophytic. After infection, the fungus leaves purple tinge on mature leaves.

Epidemiology : The disease is influenced by high humidity and defused light. The host parts which are devoid of surface free water get infected soon. Hence, cloudy and cool nights are most congenial for the disease to develop fast. The water shoots coming out under the tree canopy are the first to get infected.

Control Measures: The disease is easily controlled by dusting the foliage with finely powdered sulphur (Devarajan 1943; Ramakrishan 1954), and spraying of cosan, thiovit and calixin (Ram et al. 1978; Ram and Naidu 1978). Prophylactic spray should be at a proper time to achieve effective control.

Scab (Elsinoe faweetti Bitancourt)

It is a common disease of sour orange, Rough lemon and Tangelo, and has been reported from West Bengal, Assam, Uttar Pradesh, Andhra Pradesh and Karnataka and is mostly prevalent in the nursery. It is also recorded on sweet oranges in Karnataka (Rawal and Ullasa 1988). About 25-55 per cent of the fruits are completely spoiled in Assam (Chowdhuri 1955) by scab.

Symptoms: It rarely leads to the leaf drop but foliage distortion by the buff coloured warts may suggest thinness of the top, especialy in nursery. On leaves, small, semi-translucent dots appear usually on the underside of young leaf that become well defined pustules. The lesions mostly appear singly on one side of leaf. A few days later, these projections develop a cream to yellow-orange colour at the tips. The opposite surface corresponding to warty growth shows a circular depression with a pink to red depression. Severely attacked leaves become distorted and occasionally shed prematurely. The lesions may also appear on shoots. Sprouts of trees and the stem of nursery plants develop scab lesions similar to those on leaves.

Other parts of the tree on which scab lesions may be found are tender twigs, blossom pedicels and buttons. Eruptions occur either as individual round pimples or on the confluence of many infections. Heavily infected fruits may drop shortly after being attacked and those remaining on the trees may be scarred and distorted. Scab sometimes may be confused with similar effects produced by the other agencies. It does not invade the flesh. The lesions on fruits have corky projections which often break into scabs. When fully formed scab lesions are raised and are buff to pink to drab in colour (depending on the stage of infection).

Epidemiology : The fungus produces mostly the conidial stage (Sphaceloma fawcetti var. scabiosa Jenkins) on the host. The pathogen perpetuates and survives in off season as ascospores. Conidia are formed from 7-30°C with 66-100 per cent humidity. The fungus can infect tissue only when the surface is wet, but prefers 16-23°C temperature (Pathak 1980). Its spread is through rains. In India, the disease is a problem in areas with low temperature and high humidity. The water is the most important factors influencing severity of the disease (Yamada 1961) and can infect fruits in summer as well as in spring.

Host Resistance: The varieties/lines reported to be resistant against this disease are Sohymndong, Milan, Khatta Jamir, Brazillian Rough Lemon and Chase Rough Lemon (Rawal 1990).

Control Measures: Spraying of bordeaux mixture has been reported by Kar and Saha (1943), and Chowdhuri (1953). Spraying with difolatan (0.2%) also gives effective control (Rawal 1990). Ferbam 7.5 lb 75 WP or fixed copper have been recommended elsewhere. However, Ferbam has been found to be superior over copper spray (Knorr 1973).

Twig Blight (Colletotrichum gloeosporioides Penz.)

It is also characterized as die-back of twigs in large numbers preceding by the shedding of leaves and is present wherever citrus is grown. This disease is also caused by the association of Fusarium spp. and Diplodia natalensis Pole.

Symptoms : Blight is characterized by three types of symptoms. Wilting of foliage, die-back and post-emergence decline of watershoots. The first indication of blighted trees are dulling of the foliage followed by wilting. Symptoms may appear throughout the conopy. Gradually the leaves roll tightly, loose turgidity and drop. Die-back sets in after the leaves are shed. Sometimes the leaves shed due to the attack of powdery mildew and leaf rot pathogens. Such twigs also show die-back symptoms. Successive flushes in affected portions of the trees are retarded and dwarf or fail to develop. Such affected trees often bloom heavily, however, fruit set is poor.

Epidemiology : The disease has been found to be predisposed by the attack of powdery mildew and other leaf spot and rot causing agents. Adverse soil pH and deficiency of zinc, manganese and boron have also been suggested by Childs (1953). However, the C. gloeosporioides has also been seen to cause the vascular obstruction. Many times the insect infestation also led to the fungal infection. Though the disease appears in both spring and rainy season, it is more prevalant in the rainy season (Sohi and Kapoor 1990). The soil pulverization and adequate fertilization and maintenance of adequate soil moisture is very important. Spraying of proper insecticide is also recommended to control the insect attack. Pruning of the diseased twigs followed by the spraying of bavistin (0.1%) or captafol (0.2%) 3 times at monthly intervals have been found to be the most effective (Rawal 1990).

Alternaria Leaf Spot Fruit Rot (Alternaria citri Ellis & Pierce)

An important foliage disease in the nursery, particularly on Rough lemon and Rangpur lime.

Symptoms : Lesions develop following infection of young leaves and appear within the leaf blade or along the margins. They are regular to irregular spots surrounded by chlorotic halo, may coalesce to involve much of the leaf area. The colour of the necrotic tissue ranges from light to dark brown, edges being darker. Severe attacks lead to the leaf drop. This also attacks fruit.

A decay of fruits on trees usually is visible externally, but sometimes the decay is observed when the fruits are cut opened. Juice of a few fruits with black rot may taint an entire box. The most conspicuous symptoms appear a few weeks after colour breaking stage. Affected fruits develop an orange colour prematurely. The infected fruits usually show light brown to blackish discolouration of the rind near the styler end. Internal symptoms consist of a brown to greenish black discolouration and disorganization of the core, more clear at the styler end. Even a little infection leads to the bitter flavour to the flesh. Affected fruits are ultimately shed.

Control Measures : Spraying of copper oxychloride fungicides (0.3%) or Dithane Z-78 or chlorothalonil (0.2%) give a very effective control. Infected fruits may be collected and destroyed.

There are other diseases, namely, wilt (Fusarium spp.), dry root rot (Macrophomina phaseolina), Diplodia gummosis (Diplodia netalensis), pink disease (Botryobasidium salmonicolor), malanose (Phomopsis citri) and other root rots due to Ganoderma lucidium and Armillariella mellea which frequently occur in the fields and need attention.

Citrus Canker : The disease was first recorded in Japan. Now it is known to occur in every citrus growing area of the world.

Symptoms : It may effect any part of the tree above the ground. The leaves, twigs, young branches and fruits are the most susceptible. It first appears as small watery, transluscent spots, usually dark green in colour than the surrounding tissue and with a raised convex surface. As disease advances, the surface of the spots becomes white or greyish and finally ruptures, exposing a light brown, spongy, central mass developed in crator like formation. The spots usually surrounded by a halo. On young twigs, lesions are similar to those on leaves and fruits but on older twigs they are irregular in shape. Fruit lesions are not surrounded by yellowish halo and the crater like appearance is more pronounced. In severe cases, fruits may crack and drop.

Grape fruits, sweet oranges lime and lemon are very much susceptible to canker. Casual organism is Xanthomonas campestris pv vitri (Hasse) Dye.

Disease Cycle : The pathogen survives on diseased twigs and leaves. It enters the host through various types of wounds, natural openings, etc. On penetration in the host, it multiplies in the intercellular spaces, dissolves the middle lamella and establishes itself in the cortex. The disease is chiefly disseminated by wind, rain and insects. Citrus leaf miners are very important in the dissemination of this disease. However, man seems to be the chief agent of dissemination.

Control : It is mainly attained through: (1) quarantine, (2) eradication and destruction of infected trees, (3) spraying of neem cake at the rate of 7 kg per acre have been reported to be highly effective in checking citrus canker as well as leaf minor, (4) spraying of streptomycin sulphate (0.05%) at 75 days interval. Phytomycine (0.02%) has also been reported to be effective in checking the disease, and (5) spraying with copper oxychloride or bordeaux mixture (1%) along with pruning of infected plant parts were found effective.

Jackfruit

Jackfruit (Artocarpus heterophyllus) is consumed as fruit as well as vegetable. The trees are supposed to be hardy but even then the following diseases become a limiting factor in its cultivation.

Leaf Spot (Colletotrichum gloeosporioides)

It is most common disease occurring on Jackfruit. The disease is characterized as dark brown to brick red spot on both the leaf surfaces. These spots later turn into greyish white centres with dark brown boundaries. Dark coloured acervuli of the fungus are observed on greyish region which, when torn, ooze out profuse spore masses. The fungus has been seen to infect the young tender shoots and petioles of the leaves also. The disease is effectively checked through the sprays of carbendazim (0.1%) or methyl thiophanate (0.1%) or captafal (0.2%) or chlorothalonil (0.2%) (Rawal and Saxena 1997).

Die-back (Botryodiplodia theobromae)

It is a destructive disease of Jackfruit. The onset of die-back becomes evident by discolouration and darkening of the back some distance from the tip. The dark area advances and young green twigs start withering first at the base and then extending outwards along the veins of leaf edges. The affected leaves turn brown and their margins roll upwards. At this stage, the twigs or branches die, shrivel and fall, and there may be an exudation of gum from affected branches. Such branches have also been found to be affected by shoot borers and shot hole borers.

The infected twigs show internal discolouration when split open. In early stages, epidermal and sub-epidermal cells of twigs are often slightly shrivelled. On such twigs, erumpent aceruvuli of Colletotrichum gloeosporioides are also observed. To check this disease, pruning of infected twigs followed by spraying of carbendazim (0.1%) or Topsin M (0.1%) or chlorothalonil (0.2%) have been found effective (Rawal and Saxena 1997). The control of shoot borers and shot hole borers by spraying suitable insecticides is also recommended.

Pink Disease (Botryobasidium solmonicolor)

It is widespread in tropical and subtropical areas. The disease appears as a pinkish powdery coating on the stem. Pink colour represents profuse conidial production of fungus. Young woody branches of the affected trees loose their leaves and show die-back. Similar disease has also been recorded on mango. The affected branches should be pruned and the cut end should be pasted with bordeaux paste or copper oxychloride (Pathak 1980).

The leaf spot due to Phyllosticta artocarpina has been reported but does not seems to be serious on Jackfruit.

Mangosteen

Garcinia mangostana L. is the popular fruit of South-East Asia. Though not much work has been done as regards the pathological problems occurring on mangosteen, however, a few diseases have been reported in the literature.

Thread boll caused by Pellixularia koleroga Cool has been reported to be the common disease in Puerto Rico. This fungus also goes to coffee, cocoa, rubber trees and many other plants and is common under high humid conditions. This disease first appears as a filmental growth on young stem which gradually goes on young leaves and starts attacking from the margins where it forms the white mat like structures. The immature fruits are susceptible to this disease which become mummified and are unfit for consumption.

The affected leaves begin to loose their luster, turn brown, dark brown and finally black leading to their fall. The leaves have a typical web like structures on them. The disease has been reported to be checked by spraying bordeaux mixture (5:5:30) or copper oxychlorides. Wilt (Zignoella garcineal P.) is a common disease in Malaysia which causes the formation of tubers on the new stems that are later seen as mature tubers of the older stem. This results in the wilting and death of the plant which affects first leaves, then branches and later the entire tree. The best method of control for this diseases is to cut and destroy such branches well in advance to avoid further spread of the disease.

Rambutan

Rambutan (Nephelium lappaceum L.) is one of the best known fruits of South-East Asia (Muchjajib 1990; Tindal 1993). The trees are severely attacked by powdery mildew caused by Oidium nephelii. It attacks leaves, flowers and young shoots. The affected flowers drop off leading to failure of the crop. The disease is favoured by the dry humid weather. The disease can be best checked by spraying wettable sulphur, bayleton or systhane fungicides.

Fruit rot due to Phytophthora necotianae var. parasitica and P. botryosa are recorded to be serious in most areas. Anthracnose (Colletotrichum sp.) causes severe damage on leaves, inflorescence and young fruits. Stem canker (Fames lignosus) has been reported to be serious in the Philippines. Phomopsis leaf spot, stem canker (Dolabra sp.) and stem die-back (Thyronectrichia pseudotria) are recorded to be the minor diseases. Then there are post-harvest diseases caused by various fungi like Aspergillus spp., Botryodiplodia theobromae Port, Colletotrichum spp., Fusarium spp., Pestalotia spp. and Phomopsis homopsis. Most of these are also pathogenic in the field.

Litchi

Litchi chinensis Sonn. as such does not have much of a disease problem (Galan Sanco 1989, 1991). However, tar leaf spots caused by Pestalotia pauciseta Sacc., Botryodiplodia theobromae Pat. and Colletotrichum gloeosporioides occur in litchi orchards in some areas. The leaves occasionally showed mixed infections of B. theobromae, and C. gloeosporioides and some spots may reveal the presence of P. pauciseta. The lesions coalesced at the later stages. Older leaves show higher percentage of infections than the younger leaves. Generally, the incubation period of P. pauciseta, B. theobromae and C. gloeosporioides ranges from 5-10 day, 3-7 days and 4-8 days respectively. The diseased leaves serve as the source of primary infection. The leaf spots spread rapidly during humid and rainy months. The fungicides like bavistin, Topsin M, baycor and dithiocarbmates will control the disease.

There is another serious leaf spot of litchi incited by Microdiplodia litchi Pathak and Desai that has been recorded at Udaipur, Rajasthan. The diseased leaves show yellow brown to brick red areas mostly towards margins. The coloured areas gradually become light brown and show black dot like pycnidia. The other problems on litchi as listed are due to Clitocybe tabescence. Sudden wilt caused by Fusarium sp., die-back due to Phomopsis and foliar necrosis by Gloeosporium spp. are reported. There are some of the post-harvest diseases which are carried along with the fruits during transit and storage. Among them, the pathogens like B. theobromae, C. gloeosporioides, Cylindrocorpon forkinense, and Phomopsis spp. and Pestalotia species are the most important and need attention as they are causing diseases on the plants also.

10

Harvesting and Yield

Harvesting

Apple

Maturity Standards : In apple, the number of days taken from full bloom to harvest are 132-134 days for Starking Delicious, 138-140 days for Red Delicious and 147-148 days for Golden Delicious varieties in the high hills (above 2000 m) of H.P. In J&K State, the variety Maharaji requires 160 ± 3 days after full bloom for proper harvest maturity. Apple cultivar Granny Smith takes 180 ± 5 days for maturity.

Maturity Indices for Commercial Cultivars of Apple

Cultivar	*Days from full bloom*	*Firmness (kg)*	*TSS (%)*
Royal Delicious	125 ± 5	8.2 ± 0.40	13.0 - 15.0
Red Gold	122 ± 3	8.3 ± 0.20	12.0 - 13.5
Red Delicious	134 ± 5	8.4 ± 0.40	10.0 - 14.0
McIntosh	135 ± 4	6.8 ± 0.25	11.5 - 13.5
Golden Delicious	148 ± 6	8.4 ± 0.40	12.0 - 14.5
Granny Smith	180 ± 5	8.7 ± 0.30	11.5 - 13.0

Fruit Drop Control : Pre-harvest fruit drop in apple could be effectively checked with the application of 10 ppm NAA 20-25 days before harvest. Pre-harvest fruit drop is quite serious in early ripening cultivars like Tydeman's Early, Red Gold and Pippins, where loss of 40-60% crop load is often experienced. Mid-season Delicious group cultivars also experience 15-20% loss.

Colour Development and Enhancement of Ripening : In apple, fruit colour development in warmer and lower (below

1800 m) elevations is generally poor. Application of Ethephon (2-chloroethyl phosphonic acid) about 10 days before harvest improves red coloration. Induction of early maturity by 7-10 days in mid hills situations could be achieved through an application of 500 ppm of ethephon + 10 ppm of NAA 3 weeks before harvesting. In H.P, two pre-harvest sprays of 0.5% $CaCl_2$, followed by a 0.03% surfactant like Tween 20 for 1-2 minutes have been recommended for better shelf-life of apple.

Fruit Thinning : In apple, for obtaining better grade fruits of optimum size a post-bloom spray of 100-200 ppm ethephon 2-3 weeks after full bloom has been recommended both in H.P. and J&K.

Pear : Harvesting time of pear is determined by slight change in fruit colour and easy separation of fruits. Fruits should not be allowed to ripen to soft condition on the tree and should be harvested when they are still hard. Depending on cultivars fruit may be ready for harvest from 70 days after full bloom (viz. cv. China pear) to 135 days after full bloom (viz. cv. Barlett). The Bartlett pear normally takes 122 ± 3 days in valley (Kuala) areas of H.P.

Cherry : Sweet cherry should be harvested when fruit stalk separates easily from the spur and taste develops typical to the cultivar. From full bloom it takes 45 days in cultivars like Guigne Noir and Gross Lucenta, whereas in cultivars like Bigarrean Napoleon the number of days varies from 65-75 days.

Peach : For harvesting peach at the right time, the proper colour development in fruits and pit browning are considered as reliable guides. In yellow fleshed cultivars, deep orange colour development on fruits is associated with proper maturity. At least 5% of the pit area also should be brown if the fruit is to develop good flavor at ripening. In the peach variety July Elberta, it takes about 90-95 days after full bloom, whereas on cv. Elberta it normally takes 100-105 days. In low chilling cultivars like Flordasun, fruits should be harvested at 50% colour development stage.

Plum : In free-stone cultivars of plum, pit browning up to 5% of the pit area is a good index of maturity. For proper fruit maturity, early cultivars like Sharp Early and Formosa take 95 to 105 days, mid season variety Santa Rosa requires 115-120 days and late cultivars like Grand Duke take 130-135 days after full bloom.

Apricot : Apricot fruits of a single tree ripen for a period of about 3 weeks. The fruits develop detectable flavor when allowed to ripen on the tree. The early cultivar Charmagz matures within 75 days from full bloom, while the late cultivar Quetta requires 125-130 days for proper harvest maturity.

Yields

There is a wide range of variation in fruit yield in different States and in different cultivars. The average yield of apple and pear have been indicated earlier. Although average productivity of apple in H.P. has been shown as 9.4 t/ha, some well managed orchards of Delicious apple in Kotgarh area of the same State yield about 50 t/ha. Similarly, yield record of 35 t/ha has been achieved in 12 year old plantations of coloured mutant apple cultivars on MM 106 rootstocks under high density planting system (2222 plants/ha) in a research station in Shimla hills of H.P.

Average Yield of Different Stone Fruits by State

State		*Fruit Yield, T/ha*		
	Peach	*Plum*	*Apricot*	*Cherry*
H.P.	0.50	1.31	0.42	-
J&K	0.73	0.74	0.20	1.73
U.P.	2.11	2.50	0.28	-
NEH Region	2.50	1.18	-	-

Marketing

The biggest wholesale market for apple is the Fruit and Vegetable market at Azadpur, in Delhi. About 70 per cent of the total trade of apple is distributed through this market. The seasons of market arrivals from North Western States are as follows:

H.P. - July to October with a peak in August - September

J&K - August to November, with a peak in September - October

U.P. - June to October, with a peak in July - mid September.

There are a number of marketing channels, of which the predominant are:

(a) Farmer—Pre-harvest contractor - Commission agent-Wholesaler - Retailer-Consumer.

(b) Farmer—Forwarding agent - Commission Agent-Wholesaler - Retailer-Consumer.

(c) Farmer—Commission agent-Wholesaler-Retailer-Consumer.

The sale through pre-harvest contractors is the most important system of marketing. Normally, the small orchardists sell their crop at flowering stage to contractors who organize plant protection practices, picking and packaging of fruits.

The medium and large orchardists prefer to market their produce through channels (b) and (c). In H.P., about 65% of total apple produced is marketed through these two channels. Some Growers' Cooperatives and Government controlled marketing Corporations (like HPMC in H.P.) also get involved in the apple market. For example, in H.P the volume of trading handled by Growers' Cooperatives is about 3.5 to 4.0 per cent, while another 2-2.5 per cent of the total produce is marketed by HPMC.

The National Horticulture Board (NHB) of the Government of India regularly publishes and announces the wholesale price and market arrival figures of apple in different terminal markets for the benefit of the growers. Some quantities of apple are placed in cold storage facilities.

The Himachal Pradesh Horticultural Produce Marketing and Processing Corporation Ltd. (HPMC) has set up modern fruit packing houses in different parts of the State. Each packing house has a capacity to grade and pack 55,000 to 275,000 cartons of fruits.

Cold storage facilities at the production centres and terminal markets at Delhi, Bombay (Mumbai), Madras (Chennai) and Calcutta have been created. Similarly, the HPMC has undertaken large scale processing of temperate fruits like apples, peaches and others.

In H.P, grading and packing of fruits have been organized reasonably well. In apple, 7 size grades have been prescribed and dimensions of packing boxes, size of wrapping papers and number of layers of fruits for each grade have been standardized. In apricot, plum and peach, three size grades have been prescribed. Presently CFB (Corrugated Fiber Board) telescopic tray pack cartons for apple and smaller size universal CFB cartons for stone fruits are in use. These are available with HPMC at their various packing stations. Plastic crates (both collapsible and non-collapsible) are also used for cold storage and for processing units.

Picking and packaging of fruits are done in different ways at different sites. In U.P hills, trees are generally strip picked into small wicker baskets or small gunny bags which are emptied into larger baskets for carrying to packing stations. Both U.P. and J&K have developed packaging units for different temperate fruits.

India presently exports a small quantity of apple (5.95% of total fresh fruit export), mostly to Bangladesh and Sri Lanka. Apple cultivars like Red and Royal Delicious, Ambri and other new coloured cultivars are suitable for the export market. According to the Trade Year statistics of 1993, India exported only 7000 tons of apple during that year.

Processing

The HPMC in Himachal Pradesh owns two fruit processing plants with a combined capacity of 30,000 tons annually. The plants are equipped with modern machinery for six-fold juice concentration and aroma recovery. The plants have also facilities for aseptic bulk packaging and pasteurization. The HPMC manufactures mainly primary products like fruit pulps, single strength juices and juice concentrates. It covers fruits like apple, pear, peach apricot and plum. The pomace of the fruits after juice extraction is used for cattle feed, extraction of pectin etc.

In J&K, low grade fruits are processed for various products like apple juice, jam, jelly, fruit bars, rings, vinegar etc. from pome and stone fruits. In U.P. hills also, various types of processed products are prepared from temperate fruits. The units are mainly small and medium in size and not to the scale of HPMC. On the research side, important cultivars of apple were screened for their suitability for preparation of juice, jam, dehydrated rings, chops, cubes and preserves. Golden Delicious cultivar was found to be the multi-use cultivar to meet the FPO quality requirements. A new process for canning of apples called osmo-canning has been developed.

Similarly, process for preparation of cider, wine and vermouth have been developed. Utilization of apple pomace for production of pectin and natural colours have been attempted. Although CA storage concept has not yet become popularized, fruits placed in CA store at Shalimar in J&K have shown that this system of storage has great promise for temperate fruits.

Potential for Deciduous Fruits

Area Expansion

Both North-West and North-Eastern regions of India offer large areas ideally suitable for cultivation of pome and stone fruits. In the North-West India, stone fruits like peach, plum and apricot come up well at elevations between 900 to 1500m with an annual rainfall of 90-100cm. Apple, cherry and pear are commercially successful at elevations between 1500-2700 m above msl. The cold arid regions between 1550-3650m with annual rainfall of 25-40 cm are again highly suitable for dry nuts and drying type of apricots. There are vast tracts of land still available for further expansion of these crops.

In the North-Eastern Hills, excepting the rain shadow belts of Arunachal Pradesh and high altitude Lachung area of Sikkim, apple may not be successful commercially. Other crops like pear, peach, plum and apricot offer good scope for further expansion.

Widening the Cultivar Base

The low productivity and poor quality of apple are often linked with monoculture of a few old cultivars and their degeneration over the years. The U.P hills, particularly Kumaon division, has the unique advantage of harvesting apple fruits for early market. Similarly, the rainshadow belts in North-Eastern Hills can offer good quality apple for the eastern Indian markets, thus reducing the cost of long distance transportation from North-Western Hills. Part of the markets of Bangladesh can be captured by the fruits of the North-East.

There is good scope for introduction of new promising cultivars, replacing the Delicious group. Similarly, use of clonal rootstocks of Malling and Malling-Merton series and even their indexed material 'EMLA' selections will greatly change the productivity and quality of fruits. High density planting with spur type cultivars offer good scope. The identification of low chilling peach, plum and pear cultivars offer good possibilities for their cultivation in the low hills and in sub-tropical plains. Some of the new hybrids, including scab resistant apple cultivars need verification trials on a commercial scale.

Management Practices

Scientific water management and practicing proper training and pruning of trees including introduction of renewal pruning techniques, will make significant impact on increased production even in the existing orchards. Drip irrigation, in-situ water harvesting and correcting macro and micro-nutrient deficiencies will go a long way in bringing notable improvement in productivity as well as fruit quality. By adopting IPM strategy and organic farming practices, selected export markets can be targeted well.

Processed Products

There is immense scope for increasing various processed products of pome and stone fruits, for which technologies are available. The existing capacity of the most organized processing unit of HPMC is only marginal as compared to the volume of fruit available for processing. The HPMC utilized only 1 to 1.5 percent of total cull fruit available. The present combined capacity of two units of HPMC is 30,000 tons annually, which can be easily raised to 50 to 75 thousand tons.

The expansion of grading and packing stations, their further modernization with mechanical grading equipment, use of CFB boxes and more number of pre-cooling and cold storage units will improve the marketing system and enhance marketability of the produce. Apple has been identified as one of 6 most promising fruits for fresh fruit exports.

Constraints in Deciduous Fruit Production

Large number of old orchards (more than 30 years old) are showing decline in terms of growth and fruit yield. Such old trees do not produce adequate extension growth. Large scale replanting is therefore needed.

Delicious group of cultivars constitute the major share (about 83% in H.P.) of apple production in the country. These cultivars are self unfruitful and need cross pollination to ensure good fruit set. Interplanting pollinizer cultivars (Golden Delicious, Jonathan, Red Gold, Lord Lambourne etc.) in the proportion of 25 to 33 percent is necessary for good fruit set, and choice of wrong pollinizers and their inadequacy in number often result to low productivity.

In many countries, Delicious group has been replaced or is in the process of replacement with more promising cultivars. The need for injecting new blood into the apple industry through spread of new cultivars (spur types, colour mutants, strains of Gala, Red Fuji; scab resistant cultivars, bud sport selections of Royal Delicious, and some of the promising hybrids) is urgently felt. Some of the spur type and coloured mutants are already popular with farmers and high density planting has also caught the imagination of developmental departments and agencies both in H.P. and J&K. The research system has already identified Early, Mid and Late cultivars for different agro-climatic regions.

The low chilling cultivars of stone fruits have also covered large tracts of the sub-tropical plains of Punjab, U.P and H.P. For the hills, promising cultivars identified need further spread.

Generally, apple is grown in marginal land and fertilizers are not applied according to the requirements of the trees. The water and fertilizer use efficiency is generally poor. Also, spring frost and hailstorms are adverse weather parameters leading to low productivity. Research results have shown that through proper orchard management practices (soil and water conservation and fertilizer application) the fruit yield can be doubled in the existing orchards. The adoption of improved production technology developed by the research system can bring visible and perceptible changes in the temperate fruit industry in India. Technologies like use of clonal rootstocks, introduction of renewal pruning techniques and micro nutrient applications have not been transferred and adopted at a satisfactory level.

Apple scab disease has been the major plant protection problem in apple in J&K and H.P, whilst U.P hills are comparatively free from the disease. Apple scab forecasting system developed and the chemical control schedule prescribed have helped in reducing loss due to apple scab to a considerable extent. Apple growers are adopting the prescribed schedule of chemical sprays to control the disease. For checking entries of diseased material in the free areas of U.P. and North-Eastern Hills, strict quarantine and selection of elite disease-free mother plants are very essential.

Often it is not followed strictly. Some of the virus diseases have also been reported in apple for which biological and serological indexing/detection procedures have been developed. Limited quantity of virus-free budwood is also being supplied. Extreme care is now required to be taken to multiply quality planting material (in apple alone approximately 2 million plants/ year) for establishing new plantations.

Most of the orchardists still sell their crop at flowering to contractors as there is no well organized marketing system. Transportation in the hills itself is problematic. Post-harvest management problems originating from poor harvesting (strip picking) and improper packing system (non CFB boxes) and lack of proper pre-cooling and cold storage facilities result in huge (25-30%) loss of fruits.

Capacity of the processing sector is also inadequate. Product diversification, value addition and market infrastructure development would require very substantial investment. The existing processing units are quite old and they require modernization for which substantial investment is required. CA storage trials have shown good promise. Its extension in larger growing areas is needed. Technology for storage of apple is now known, as a result of which apple is now available throughout the year.

Government Policies and Plans

Research on pome and stone fruits is conducted mainly by three State Agricultural Universities, namely: (a) Sher-e-Kashmir University for Agriculture Science & Technology, J&K; (b) Y.S. Parmar University for Horticulture and Forestry, H.P.; and c) G.B. Pant University for Agriculture & Technology, U.P. A good number of research stations of these Universities located in major pome and stone fruit growing belts are engaged in temperate fruit research.

The Indian Council of Agricultural Research (ICAR) through All India Coordinated Research Projects on Fruits and Post-harvest Technology and another Network project on Apple Scab disease has further strengthened the research activities in deciduous fruits. A few long established temperate fruit research stations namely at Shalimar Bagh in J&K, at Mashoobra in H.P. and at Chaubatia in U.P. hills have made commendable progress in temperate fruits. Both State Governments and the ICAR provide financial support to strengthen fruit research in these Universities. A large number of ad-hoc research projects on pome and stone fruits funded by ICAR are also generating good information on these crops.

During the 8th Government Developmental Plan (1992-1997) the ICAR established a Central Institute for Temperate Horticulture (CITH) with its headquarters at Srinagar in the J&K State, with a regional station at Mukteshwar in the U.P. Hills. Both these research stations will work exclusively on temperate fruit crops. This new Institute will receive major support during the 9th Plan period (1997-2001). In the North-Eastern Hills region, the ICAR Research Complex for NEH Region with its headquarters in Meghalaya State and regional units in each of the 5 other States are engaged in fruit research.

On the developmental side the State governments are engaged in nursery production of quality planting material. For example, in H.P. alone currently there are 600 nurseries in private and public sector producing and distributing more than 0.8 million plants of apple alone, every year. There are ambitious programmes in all the States to further expand/ replant with new improved cultivars. Apple scab disease control and post-harvest processing sectors are getting focused attention in Government developmental plans.

The Directorate of Marketing and Inspection of the Government of India has framed grade standards for apple, plum and William pear. The organizations like National Horticultural Board (NHB), National Cooperative Development Corporation (NCDC), Agricultural and Processed Food Products Export Development Authority (APEDA) etc., are providing incentives to traders and exporters to improve their infrastructural facilities like grading and packaging centres, refrigerated transport, setting up of pre-cooling, cold storage, auction platforms etc. The NCDC is undertaking procurement and marketing of apple on a limited scale. The NHB has set up a market information service for the benefit of growers.

Deciduous fruits, covering pome and stone fruits contribute significantly to the horticulture economy of India. Apple production dominates the scene and systematic cultivation and marketing of apple can change the rural economy in the hills of North-Western India. New vision and concerted efforts are required for change in variety mix, supply of quality planting material from elite clones on indexed clonal rootstocks. High density planting, water management including micro-irrigation, integrated plant nutrient management and IPM strategy for plant protection are some of the areas which need greater R&D focus. Adoption of post-harvest management practices and infrastructure development for grading, packaging, pre-cooling and storage of the produce needs focused developmental attention. Value addition and export promotion, particularly of apple are drawing due attention of the developmental agencies in India.

Fruit Culture

Introduction: The Cashewnut (Anacardium occidentale L.) is one of the important nut tree grown in the tropical world.

This fruit tree was introducted into Malbar coast of India in 16th century by Portuguese. It is the native of Brazil. It is grown mostly on West Coast area of Maharashtra, Goa, Karnataka, Kerala and Tamil Nadu.

Importance: The dry nuts - Kernels are of high nutritive value and also rich in protein, carbohydrates, minerals like calcium, phosphorus and iron. This has a great export value and is considered as one of the important horticultural crop in India.

Climate: The Cashew requires a minimum rainfall of 60 cm per annum, but can stand extremes of rainfall from 20 cm to 400 cm. If there is sufficient water supply, it can withstand a long period of dry spell and low humidity. The Cashew is a sun loving tree and does not tolerate excessive shade, also it does not favour very high temperatures above 45 deg cent. During the fruit set and development. Heavy rains and cloudy weather during flowering adversely affect the yield.

Soil: The Cashew is cultivated on a wide variety of soils. It is considered a crop of marginal land and is recommended for slopy and light soils. The best soils for better production are deep, friable, well drained and without a hard pan upto 2/3 m in depth. The Cashew is mainly grown on Laterite, red and coastal sands in India.

Varieties: The following are the important varieties of Cashew grown in different parts of the country.

i. Vengurla 1
ii. Vengurla 2
iii. Vengurla 3
iv. Vengurla 4
v. Vengurla 5
vi. Vengurla 6
vii. Bhubaneshwar
viii. Kanaka
ix. Dhana
x. Selection

Propagation: The cashew is grown by the following methods:

i. *Seed Propagation:* It is the oldest and cheapest method of propagation. It is also used to raise the plants for the purpose of grafting.

ii. Layering this method is more successful in coastal and more humid areas.

iii. *Soft Wood Grafting:* This is followed both in the nursery as well as in the field of in-situ. For this purpose, veener grafting is done.

iv. *Top-working:* This method is used not for propagation but for converting inferior free into desirable one.

Planting and Season: Normally cashew grafts are planted at the spacing of 7.0 x 7.0 or 7.5 x 7.5 or 8.0 x 8.0 m on plain land or on desirable contours on slopy lands. After marking, the pits of 60x60x60 cm are dug and filled with a compost,top soil, single super phosphate and karnaj cake. This is to be well before the monsoon starts. Planting should be done during June-July *i.e.* in the beginning of monsoon.

Interculturing:

i. Removal of weeds is done once twice

ii. Intercrops are planted in interspace

iii. Some perenial forest plants and/or some seasonal crops can be taken as per the local need and situation.

Care of young orchard:

i. Gap filling for the missing plants

ii. Removal of outgrowth on stocks in case of grafted plants

iii. Staking with bamboos.

iv. Cover cropping

v. Providing protective irrigation during first few summers. These are some of the points for young plantations.

Special Horticultural practices:

i. Pruning of dead and dried shoots alongwith crisscross branches and water shoots

ii. Pruning of leader shoots in June followed by 2% KNO_3 spray

iii. Spraying with 10 ppm NAA twice during flowering for increasing fruit set and minimizing flower and fruit drop.

Irrigation: The cashew is mostly grown as the rainfed crop and requires no irrigation in the high rainfall areas. However, if the rainfall is low, the cashew responses well to irrigation water at the time of fruit set upto full development stage of nuts. Irrigation should not be given before or at the time of flowering, as it is likely to promote vegetative growth.

Nutrition: To improve the growth and yield the cashew trees should be manured and fertilized with bio-fertilizers like biomeal and NpK 500,200, 200 gm per tree per year. Spraying with ultrazyme and 8:12:24:4 NPKmg twice, during fruit development is useful to increase size and yield of nuts.

Plant Protection: The crop particularly blossom and fruits should be protected against Tea Mosquito, Stem and root borer is also an important pest. Leaf minor, Leaf and blossom webber, Flower thrips are the common pests in neglected orchards. Inflorescence blight, pink disease, and anthrac nose are some diseases found on cashew. Clean cultivation, pruning, better aeration and preventive measures should be adopted besides occassional sprays of selected pesticides and fungicides.

Harvesting and Yield: The Cashew comes to its full bearing stage at the age of 10th year and continuous to give yield further for 30-40 years. Usually the nuts are collected after they fall off from the tree. They are separated from apples and dried for 2/3 days to reduce the moisture. Regarding yield there is a wide variation and differences of opinions. However, the desirable variety at good management yields 50 kg of raw nuts.

Post Harvest Handling: The processing of raw nuts involves roasting, shelling, drying, peeling, grading and packing `A' grade nuts are always in demand from the outside markets and fetch a good price. A Fenny like drinks are prepared from the apples.

The Cashewnut

The Cashewnut (Anacardium occidentale L.) is one of the important nut tree grown in the tropical world. This fruit tree was introducted into Malbar coast of India in 16th century by Portuguese. It is the native of Brazil. It is grown mostly on West Coast area of Maharashtra, Goa, Karnataka, Kerala and Tamil Nadu.

Importance: The dry nuts - Kernels are of high nutritive value and also rich in protein, carbohydrates, minerals like calcium, phosphorus and iron. This has a great export value and is considered as one of the important horticultural crop in India.

Climate: The Cashew requires a minimum rainfall of 60 cm per annum, but can stand extremes of rainfall from 20 cm to 400 cm. If there is sufficient water supply, it can withstand a long period of dry spell and low humidity.

The Cashew is a sun loving tree and does not tolerate excessive shade, also it does not favour very high temperatures above 45 deg cent. During the fruit set and development. Heavy rains and cloudy weather during flowering adversely affect the yield.

Soil: The Cashew is cultivated on a wide variety of soils. It is considered a crop of marginal land and is recommended for slopy and light soils. The best soils for better production are deep, friable, well drained and without a hard pan upto 2/3 m in depth. The Cashew is mainly grown on Laterite, red and coastal sands in India.

Varieties: The following are the important varieties of Cashew grown in different parts of the country.

i. Vengurla 1
ii. Vengurla 2
iii. Vengurla 3
iv. Vengurla 4
v. Vengurla 5
vi. Vengurla 6
vii. Bhubaneshwar
viii. Kanaka
ix. Dhana
x. Selection

Propagation: The cashew is grown by the following methods:

i. *Seed Propagation:* It is the oldest and cheapest method of propagation. It is also used to raise the plants for the purpose of grafting.

ii. Layering this method is more successful in coastal and more humid areas.

iii. *Soft Wood Grafting:* This is followed both in the nursery as well as in the field of in-situ. For this purpose, veener grafting is done.

iv. *Top-working:* This method is used not for propagation but for converting inferior free into desirable one.

Planting and Season: Normally cashew grafts are planted at the spacing of 7.0 x 7.0 or 7.5 x 7.5 or 8.0 x 8.0 m on plain land or on desirable contours on slopy lands. After marking, the pits of 60x60x60 cm are dug and filled with a compost,top soil, single super phosphate and karnaj cake. This is to be well before the monsoon starts. Planting should be done during June-July *i.e.* in the beginning of monsoon.

Interculturing:

i. Removal of weeds is done once twice

ii. Intercrops are planted in interspace

iii. Some perenial forest plants and/or some seasonal crops can be taken as per the local need and situation.

Care of young orchard:

i. Gap filling for the missing plants

ii. Removal of outgrowth on stocks in case of grafted plants

iii. Staking with bamboos.

iv. Cover cropping

v. Providing protective irrigation during first few summers. These are some of the points for young plantations.

Special Horticultural practices:

i. Pruning of dead and dried shoots alongwith crisscross branches and water shoots

ii. Pruning of leader shoots in June followed by 2% KNO_3 spray

iii. Spraying with 10 ppm NAA twice during flowering for increasing fruit set and minimizing flower and fruit drop.

1. *Irrigation:* The cashew is mostly grown as the rainfed crop and requires no irrigation in the high rainfall areas. However, if the rainfall is low, the cashew responses well to irrigation water at the time of fruit set upto full development stage of nuts. Irrigation should not be given before or at the time of flowering, as it is likely to promote vegetative growth.
2. *Nutrition:* To improve the growth and yield the cashew trees should be manured and fertilized with bio-fertilizers like biomeal and NpK 500,200, 200 gm per tree per year. Spraying with ultrazyme and 8:12:24:4 NPKmg twice, during fruit development is useful to increase size and yield of nuts.
3. *Plant Protection:* The crop particularly blossom and fruits should be protected against Tea Mosquito, Stem and root borer is also an important pest. Leaf minor, Leaf and blossom webber, Flower thrips are the common pests in neglected orchards. Inflorescence blight, pink disease, and anthrac nose are some diseases found on cashew. Clean cultivation, pruning, better aeration and preventive measures should be adopted besides occassional sprays of selected pesticides and fungicides.
4. *Harvesting and Yield:* The Cashew comes to its full bearing stage at the age of 10th year and continuous to give yield further for 30-40 years. Usually the nuts are collected after they fall off from the tree. They are separated from apples and dried for 2/3 days to reduce the moisture. Regarding yield there is a wide variation and differences of opinions. However, the desirable variety at good management yields 50 kg of raw nuts.
5. *Post Harvest Handling:* The processing of raw nuts involves roasting, shelling, drying, peeling, grading and packing 'A' grade nuts are always in demand from the outside markets and fetch a good price. A Fenny like drinks are prepared from the apples.

Tamarind

Tamarind (Tamarindus indica L) is one of the common fruit trees found grown all over India mostly under rainfed conditions, particularly in Tamil Nadu, Maharashtra, Karnataka, Andhra Pradesh, Madhya Pradesh and Orissa. It is also one of the most popular avenue trees which yield useful fruits and timber besides providing shade. Tamarind is believed to be native of Tropical Africa but now cultivated throughout South East Asia, Australia, America.

Importance: Tamarind fruit is an important condiment/ adjunct used as an acidic/ flavouring agent in the Indian cookery. India produces about 2.5 lakh tonnes of Tamarind pulp annually. Pulp powder and juice concentrates have a export potential in European countries. Tamarind pulp has an excellent keeping quality when dried properly in and cured with salt. Pulp is rich in glucose (47.7% of total sugar); D-manose (24.5%) and D-maltose (20.4%). The sour taste of the pulp is attributed to Tartaric acid (8-18%) together with malic and citric acids (2%). The fruit is good source of phosphorus, calcium and iron. Tender leaves and flowers are also edible. Tamarind seeds yield a cheap substitute for cereal starch which is in textile industry.

Climatic: Tamarind tree has a acid climatic adaptability and can be grown in humid to dry hot regions. It is very sensitive to frost. The optimum rainfall requirement is 750-1900mm, but can thrive in region with low annual rainfall of 500-750mm.

Soil: It can thrive on variety of soils, but deep loamy or alluvial soils providing optimum condition for development of its long tap roots are ideal. It can tolerate slightly saline and alkaline soils.

Varieties: There are very few well recognized tamarind varieties. However, in the recent past few seedling selections have been identified on the basis of fruit quality and yield. These are: Prathisthan from Fruit Research station, Aurangabad, Tree No 38 located at college of Agriculture, Pune, (Maharashtra) A high yielding type PKM- 1 has been developed by Coimbtore Agriculture. University (Tamil Nadu).

Propagation: Common method of propagation in vogue is by seed. However, being heterozygtes, it does not breed true to type through seed. The prolonged juvenile phase is also due to predominance of seed propogation. Seedlings Trees about 15-20 years for economic yields. Vegetatively propagated plants start yielding after 3-4 years. Vegetative methods such as approach grafting, air layering and budding have been successful. However, layers are suitable for planting under dryland conditions for lack of deep growing tap root system. In order to obtain true to type plants with reduced pre-bearing age, uniform growth and yield vegetative methods like budding and grafting be used in establishing Tamarind orchards.

Planting: In order to provide space and proper medium for development of vigourous and deep tap root system planting should be done in pits of 1 x 1 x 1m size. The pits should be dug during summer and filled in with 2 kg single superphosphate at the base and with a mixture of well decomposed Farm yard manure and top soil. About 100g 10 per cent carbanyl powder be mixed in the soil mixture to avoid the problem of termites. The planting distance may vary from 8 X 8 to 10 X 10m depending on the soil type, closer spacing in shallow soils and wider spacing in deep soils be adopted Rootstock seedlings be planted in situ during monsoon (June-July) and budded by patched method using a bud from improved variety.

Interculturing: Interculture operations like weeding, hoeing be carried out regularly. Cover crops like cowpea, horse gram etc., may be grown during rainy season to control weeds, check soil erosion and improve soil health.

Inter crops like drumstick or low growing vegetables can be grown during initial 4-5 years to get early returns.

Care of young Orchards; Young plants be provided support of bamboo stick and irrigated during dry spell particularly summer months upto 3-4 years of planting.

In the initial years ,training is very essential to develop the farm work .The plant be trained to modify leader system. Shoots appearing from root stock and weak and diseased or dried branches be punned.

Special Horticulture Practices: Tamarind is a semiforest tree and does not need any special treatment as it is adapted to a wide range of soil and agro climatic conditions.

Irrigation: Tamarind is a dry land (rainfed) crop and hence it is normally not irrigated. But young orchards needs to be irrigated, especially during dry spells and summer months. One year old plants be given 10 litres water at an interval of 6-8 days during summer which may be doubled during 2nd and 3rd year. There after irrigation may not be necessary. However if it is provided growth will be faster and better.

Nutrition: Nutritional requirement of Tamarind have not been studied so far and standardized recommendations are not available. On an adhoc basis one year old tree should receive 10 kg FYM + 100 g N + 50g P_2O_5 + 100g K_2O and the dose be increased with age. Thus, a 10 year old tree should receive 50 kg FYM +1 kg N + 500 g P_2O_5 + 1 kg K_2O. If irrigation is available additional 500 g N be given in the month of September-October.

Plant Protection: There are hardly any pests and diseases on Tamarind plant. So far no serious pests and diseases have been reported. Nor any systematic studies have been undertaken on this plant as it grows mostly under semiwild conditions. The plant has developed natural resistance against this maladies and hence requires hardly any plant protection measures.

Harvesting and Yield: Seed propagated plants start bearing 7-8 years after planting. While grafted or budded plants will start fruiting 4-5 years of planting. Fruits are harvested during January-April months. The yield varies with soil type, climate, propagation method and management practices. A well managed tree yield 300-500 kg ripe pods.

Post-harvest handling and marketing: After harvest pods are spread on ground and cut for 6-7 days. The shell and seeds as well as the fibrous material is removed and the pulp is collected, the pulp can be stored for a period of 6-12 months after properly drying in the sun. Many a times trees or entire plantation is auctioned to the traders who manage harvesting as well as marketing.

The Sapota (Chickoo)

Introduction: The Sapota (Achras Zapota L.) is a delicious fruit introduced from tropical America and first planted at Gholrad near Mumbai in 1898. It's cultivation has spread to Maharashtra, Gujarat, Karnataka and Tamil Nadu. It is evergreen, tropical fruit tree, spreading habit and lives longer upto 100 years.

Importance: The Sapota when fully ripe is delicious and is eaten as dessert fruit. The pulp is sweet and melting. Besides food values the Sapota fruits are also used in some auverda preparations.

Climate: The Sapota is a tropical fruit crop and can be grown from Sea level upto 1200 m. height. It prefers a warm and humid weather and grows in both dry and humid areas. Areas with an annual rainfall of 125-250 cm are highly suitable. The optimum temperatures ranges between 12 deg.cent. to 36 deg.Cent.

Soil requirement: The Sapota tree is a hardy perennial and evergreen tree and can be grown on a wide range of soils. Drainage is most important. There should not be a hard pan in the sub-soil. Deep and porous soils make a good growth. The Sapota can tolerate the presence of salts in the soil or in irrigation water to some extent.

Varieties: The important and widely adopted varieties are Kali Patli and Cricket Ball (Calcutta Large). The other varieties are, pili patti, Bangalore, Baramati, Dwarapudi, Chhatri, etc.

Propagation: The Sapota can be propagated by seeds by grafting or by layering. However, commercially followed method is softwood grafting on rayan seedlings. This method has replaced the earlier method called as approach grafting. The Sapato when grafted on rayon has initially slow growth but the tree lasts longer. The layered plants grow vigorously and the method is cheaper as no root stock is required.

Planting and Season: The planting of grafts is done from June to October, for which pits of 1x1x1 m at the distance of 8 x 8 or 9 x 9 or 10 x 10 m are dug well before the onset of monsoon. The pits are filled in with a well decomposed manure,

5:10:5 mixture, single super phosphate, neem cake and a mixture of micronutrients. Hundred to 120 plants are accommodated in a hectare.

Interculturing: The Sapota has a long pre-bearing age and a wider spacing allows enough interspace for growing certain crops for few seasons. The short duration fruit crops like banana, papaya, pineapple or vegetables, different types of lillies or groundnut, chilli, gram,etc. are suitably grown for some seasons. Frequent weeding or mulching is necessary for first few years.

Care of Young Orchard: A strong windbreak should be established by planting tall and thick growing trees on the windward or on all the sides of the orchard. Frequent removal of outgrowth on the stock is necessary in grafting plants. Gap filling and supporting with bamboo is to be done as early as possible. Young plants are likely to damaged due to the scorching sun, dry and hot wind and frost. Necessary protective are adopted.

Special Horticultural Practices: Training and pruning: In Sapota, a strong central stem is necessary. The Sapota, in general, has a well balanced distribution of branches and the crown assumes a uniform shape. There is no neccessity of pruning every year. All the growths those appear on the rootstock below the graft joint must be removed. After ¾ years of planting, the lowermost branches upto a height of 1 m may be removed. Similarly, overshaded and crowded branches are also removed. In Sapota, new growth and flowering occur simultaneously and it has a mixed type of bearing habit. Flowers and fruits appear in the leaf axils in the new growth and hence pruning of branches should not be done.

Irrigation: Though Sapota can tolerate drought conditions to some extent, yet it responds well to irrigation. As the tree is perennial, evergreen and almost growing cum fruiting stage every year, irrigation - whenever is necessary must be provided.

Nutrition: For healthy growth and good quality fruits manures are fertilizers should be applied regularly. The trees should be fertilized twice in a year during June and January. A well grown tree should be supplied with 100 kg FYM, 10 kg

Biomeal 2.5 kg, 5:10:5; 1 kg ormichem or any other micronutrient mixture. The foliar sprays of NpK, Mg & Zn are useful to increase the fruitset and improve the size of fruit. Nitrophoska 8:12:24:4 at the rate of 100 gm/tree has proved beneficial. Nutritional trial at Ganeshkhind, Pune it was observed that, the use of vermicompost alone and supported with foliar sprays of biocil and biok, a bumber harvest was achieved at 3rd year of application.

Plant Protection: The Sapota crop is affected by insect pests and diseases and also by some disorders. The important insect pests are as:

i. Stem borer
ii. Scale insects
iii. Leaf webber
iv. Mealy bug.
v. Leaf minor
vi. Bud eating caterpillar
vii. Barkeating caterpillar and
viii. Fruit borer.

Important diseases:

i. Leaf spot
ii. Sooty mould and
iii. Flatenned branches.

The suitable control measures are adopted. Clean and well nourished orchard is affected less. Caryaryl, Malathion, Dimithods, Bavistin, Kuman.L,etc. should be sprayed alternatively apart from following clean cultivation.

Harvesting and Yield: Sapota is a climacteric fruit and it improves in quality after harvesting but immature fruits should never be harvested. A well grown and well nourished Sapota tree yields 2500 to 3000 fruits weighing about 150 kg every year, for 50 to 60 years during its productive age.

Post Harvest Handling: Well matured fruits ripen within 3/5 days after harvest and can be stored for 10-12 days in 12 deg.Cent.temperature. Graded fruits are packed in boxes and then transported to distant markets. For local markets, the fruits are washed in water and carried in baskets or gunny bags.

The Custard Apple

Introduction : Custard Apple (Annuna squmos L.) is one of the finest fruits introduced in India from tropical America. It is also found in wild form in many parts of India. It is cultivated in Andhra Pradesh, Maharashtra, Karnataka, Bihar, Orissa, Assam, and Tamil Nadu. Besides India, it is common in China, Phillippines, Egypt and Central Africa.

Importance: It is very hardy, medium in growth and decidious in nature. The fruits are generally used as fresh, while some products or mixed fruits like custard powders, ice-creams are prepared from the fruits. Besides high nutritive value, it has also a high medicinal value. Unripe fruits, seeds, leaves and roots are considered and used in medicinal preparations.

Climate: all annonas are tropical in origin and grow well in hot and dry climate with varying degrees of difference. Custard Apple requires hot dry climate during flowering and high humidity at fruit setting. Flowering comes during hot dry climate of May but fruit setting takes place on onset of monsoon. Low humidity is harmful for pollination and fertilization. The Custard Apple withstands drought conditions cloudy weather and also when the temperatures go below 15 deg.cent. Annual rainfall of 50-80 cm is optimum, though it can withstand higher rainfall.

Soil: The Custard Apple is not very particular about soil conditions and flourishes in all types of soils like shallow, sandy, but fails to grow if the subsoil is ill drained. It can grow well in deep black soils provided they are well drained. A little salinity or acidity does not affect it but alkalinity, chlorine, poor drainage or marshy-wet lands hamper the growth and fruiting.

Varieties: The following are some of the varieties grown in different agro-climatic regions of the country.

i. Red Sitaphal

ii. Balanagar

iii. Hybrid
iv. Washington
v. Purandhar (Pune)

Propagation: The annonas are commonly propagated by seeds. Recently some researchers have developed some techniques in vegetative methods and budding can be adopted for multiplication. The seedlings of local custard apple have proved a good root stock for many improved varieties and hybrids. Seeds treated with 100 ppm for 24 hours germinate quickly and uniformly.

Planting and Season: Planting is done during rainy season. The pits of 60x60x60 cm at spacing 4x4 or 5x5 or 6x6 depending on soil type are dug prior to monsoon and filled with a good quality FYM, single super phosphate and neem or Karanj cake under dry conditions and with drip irrigation system planting at 6x4 meters has given good growth and better fruit setting.

Interculturing: For good plant growth, the weeding should be done to keep away the weeds. Intercropping with some legumes, peas, beans and merigold flowers are commonly taken by the growers. Normally, no crop is taken during winter as the plants go under rest.

Care of Young Orchard: The gap filling is to be done as early as possible. Stagnation of water during first monsoon should be taken care of, if the planting is done on imprevious soils or ill drained soils.

Special Horticultural Practices: For uniform flowering and for early flowering and also to check the flower and fruit drop and for improving fruit size, following growth regulators are used.

i. For defoliation and bringing the plants under uniform rest Ethryl at 1000 ppm is sprayed one month after the harvest of the fruits.
ii. For better and early flowering Biocil a ml per litre of water is sprayed just prior to flowering.
iii. 10 to 20 ppm NAA is sprayed just prior to flowering period to reduce the flower and fruitdrop.

iv. During fruit development 50 ppm GA + 5 ppm + 0.5 ppm cppu, foliar spray improves the fruit size and luster of the fruits.

Irrigation: In general, Custard apple is grown as a rainfed crop, and no irrigation given. However, for early and bumper harvest of the crop. Irrigation on flowering *i.e.* from May should be given till regular monsoon starts. For better flowering and fruit setting, mist sprinkling is better over flood or drip system of irrigation as it keeps to lower down the temperatures and to increase in the relative humidity.

Nutrition: Generally, no manures or fertilizers are applied to rainfed crop. However, for early and bumper harvest with a good keeping quality, the following dose is recommended to a fully grown tree. Biomeal .... 10 kg, 5:10:5 1 Kg. Ormichem micronutrient mixture 0.250 kg at the time of flowering and another dose of 10:26:26 or 19: 19: 19 mixture after fruit setting. Foliar spray with 8:12:24:4 10 gm/lt, twice during fruit development. Occassionally, zinc or iron or both deficiencies are noticed and can be taken care of by spraying chealed zinc or Ferrous.

Plant Protection: Though the crop is hardy, it suffers from the following pests-

i. Mealy bug
ii. Scale insects
iii. Fruit boring caterpillar
iv. Leaf spot, Anthrac nose, Black stone, spraying with neem oil, Meenark, and some herbal preparation is recommended.

Harvesting and Yield: The Custard apple is a climacteric fruit and harvested at the maturity state when the fruit starts to change colour from green to its varietal colour shade. Immature fruits do not ripe. Swallowing some apical buds-showing inner pulp is also an indication of maturity. A were grown tree yield above 100 fruits weighing 300 to 400 gm. The season of harvest is from August to October.

Post-harvest Handling: The fruits do not withstand cold storage as well as handling after ripening. Firm but mature fruits can be kept at 6 deg.cent. temperature for about a week but such fruits loose aroma and attractiveness and also develop some cipid taste.

The Coconut

Introduction: The coconut (cocos nucifera L.) is a Tropical fruit plant grown on a large scale in a number of countries in tropical and sub-tropical areas. The place of origin though contradictory, is believed to be somewhere in South Asia-Malaysia. In India, coconut palms are grown on the entire coastal belt. Major share goes to Kerala, Karnataka and Tamil Nadu followed by Goa, Maharashtra, Andhra Pradesh, and Orissa.

Importance: Every part of the coconut palm is useful to mankind. It supplies food, fuel, drink, oil and shelter to us. The wet kernel, desicated copra, coconut water, cream flour, oil, cake, toddy, husk, fibre, shell, coir, wood and leaves are used for one purpose or the other.

Climate: The coconut palm grows well in a equatorial climate. The long spells of hot and dry weather, severe winters and extremes of temperatures are not favourable for coconut growing. However, it thrives well upto 1000 m above the sea level. The well distributed rainfall ranging from 100 to 300 cm is favourable.

Soil: The coconut palm can grow in a wide range of soil conditions ranging from laterite, alluvial, red, sandy loam having pH range from 5.5 to 8.0. Soil should be fertile and good drainage without any hard substratum within one of the surface.

Varieties : There are number of varieties grown in different parts, which are classified as under:

a. Tall varieties:
   i. West Coast Tall
   ii. Lakshadweep
   iii. Andaman
   iv. Kappadam
   v. Laguna
   vi. San Raman

a. Dwarf varieties:
   i. Chowghat - Dwarf green,
   ii. Dwarf orange
   iii. Malayon Dwarf
   iv. Mangipod
a. Hybrid varieties
   i. T X D
   ii. D X T
a. Selections- Pratap, Banoli, etc.

Propagation: The coconut palm is commonly propagated by seeds. Since the productivity of the palm depends on the quality of the seedlings, utmost importance should be given to select the best nuts. The mother palms should be regular bearers giving annual yield of more than 100 nuts with the copra content of 150 g per nut. The mother palm should be within the age group of 20 to 40 years. These palms producing barren nuts and shedding immature nuts in large number should be discarded for the nursery purpose.

Planting and Season: The planting distance depends on the type of plant or variety, soil type and weather monoculture or associated culture or mixed culture is adopted, distance should be in such a way that the fronds (leaves) of planting adjacent palms to do not overlap when fully grown. The spacing maybe 7 to 8 m both ways for tall varieties. Well developed and well grown seedlings about 1 to 2 years old should be selected for planting. The planting should be done on onset of monsoon for which the pits of 0.75 x 0.75 x 0.75 m should be opened and filled with a mixture of compost, bone meal, super phosphate, well before the monsoon.

Interculturing: Weeding and stirring the soil around the plant should be done frequently. The interspace may be kept clean or utilised for taking other crops, seasonals or perennials.

Care of Young Plants: The young palms need proper care during the early stages of growth from transplanting upto 5th year. The seedlings should be protected from stray cattle. During summer months, shade and frequent watering must be provided to the young palms to avoid casualties. The care should be taken to see that the collar and leaf axils of the seedlings are not covered by mud, water or soil.

Special Horticulture Practices: The dwarf varieties are considered on self-pollinating while tall varieties are considered cross pollinating inter planting of mix planting be done to increase the fruit set. Honey bees are considered to be very useful for pollination and be rared in the viscinity of palm cultures.

Irrigation: The coconut palm requires large quantities of water for its normal growth and timely production. In normal conditions and in absence of rains every palm needs about 16-18 litre water daily. Regular irrigation helps to increase the fruit set, improve size and copra content. However, stagnation of water and ill drained conditions should be strictly avoided.

Nutrition: To get a good yield, palms should be fertilized regularly and with a quick frequency, Bimonthly applications are more beneficial. Every well grown palm should be applied with 25 kg. Of FYM 1 kg of 5:10:5. Micronutrients/Ormichem @ 2 kg/palm once in a year should be applied to avoid any micro-nutrient disorders.

Plant Protection: Important pests, Rhinocerous beetle, Red palm weevil, leaf eating caterpillar and cockchafer beetle are the major insect pests. Occassionally mealy bug, rat and slug also become trouble some. Bud rot and leaf rot and root wilt can cause severe damage to palms. One should be very alert to notice the symptoms of any post and take the suitable control measures. Close planting should be avoided plantation should be kept weed free and clean. The palms should be cleared off at least once in a year. Providing a good drainage and feeding the palms with balanced nutrition is useful to keep palms healthy.

Harvesting and Yield: The well developed nuts should be harvested one month earlier to full maturity. Harvesting of nuts is a skilled and risky job. 2 to 6 harvestings can be done yearly. Average 80 to 100 nuts are harvested per tree/year.

Post Harvest Handling and Sale: The nuts are hardy and can be kept for a long time. Nuts harvested at less than

9 months age will be dried spoiled or without or very thin layer of copra. Copra is removed and then processed for different purposes or dried and stored and sold as dry copra.

Tropical and Sub-tropical Fruits

The Papaya

Introduction: The papaya (carica papaya) is one of the important delicious fruit crops grown in the tropical and subtropical parts of the world. It is originated in Mexico and spread to almost all the corners of the tropical world.

Importance: The papaya is highly productive, problematic and interesting crop. It is easy to grow and short duration crop. As a raw fruit, it is used in cooking and some preparations. Its latex is used as a papain in food and medicine industry. Ripe fruit is very delicious containing vitamin A and Carbohydrates.

Climate: The papaya being tropical crop, favour high temperature and high humidity. It is very susceptible to frost and hail storm. The long days are favourable for good quality and flavour. During flowering, high rains are injurious and cause heavy damage.

Soil: The papaya grows under wide types of soils. However, very shallow and very deep black soils are not suitable medium, fertile, well drained and lime free soils are preferred for papaya cultivation.

Varieties: Important varieties under cultivation

Pusa varieties:

a. Pusa delicious
b. Pusa majesty.
c. Pusa jiant
d. Pusa dwarf
e. Pusa nanha

Coimbatore varieties:

a. Co_2
b. Co_3

c. Co_5
d. Co_4
e. Co_5
f. Co_6

Bengalore varieties:

a. Coorg honey
b. Pink flesh

Pantnagar varieties:

a. Pant 1
b. Other varieties: Ranchi, Washington, Honey dew, Taiwan varieties like, golden queen, No 815 and others.

Propagation: Commercially the papaya is propagated by seeds. The tissue culture technique is limited to research laboratories only. The seeds loose viability in a short period and therefore the seeds should not be stored for more than a season. The seedlings in polybags are prepared. Due care is taken to avoid damping off of the newly germinated and young seedlings. The seedlings become ready for transplanting within 6-8 weeks.

Planting and season: Planting is done during the flowering seasons:

a. Spring season (February - March)
b. Monsoon season (June-July)
c. Autumn season (October-November)

Heavy rains, hot air, frost, etc. are considered while selecting the season for planting in a particular area. The pits of 30x30 x 30 are prepared in already selected and prepared field at the distance of 2.5 to 3 meters distance. The pits are filled with well decomposed FYM and NPK mixtures. A care is taken not to disturb the roots while transplanting the seedlings.

Intercultivation: Interculturing is mainly done to remove the weeds during the early period of growth, weeding and hoeing in between rows also favour better aeration to root zone. Some times pre-emerging weed killers like Basalin are used. Secondly, roughing is done to remove extra plants, weak plants and affected plants. After ensuring one plant per pit, earthing up is done 30 cm in radius around the plants.

Special Horticultural Practices: Planting of 3 to 4 seedling at one pit and then removing extra plants and keeping one plant per pit while doing so 10 per cent male plants are kept in female plant population to have pollination and to improve the fruit set.

Irrigation: For better growth, production and quality, the optimum soil moisture is maintained by irrigating the crop judiciously. Irrigation interval well depends on season, crop growth and soil type. In no case, water should be allowed to stagnate causing root and stem rot. Drip system of irrigation is beneficial and the actual quantity of water to be given per plant per day should be worked out critically.

Nutrition: The papaya is a very feeder and requires the application of chemical, organic and biofertilisers. The dose of NPK @ 500 kg each per ha, along with 20-25 tonnes of FYM, 50 to 100 kg of ormichemi mirconutrients and ultrazyme sea weed extract granules 25 kg is found for a crop of about 50 tonnes within 18-20 months. Additional 60% of this dose is again applied for the second flush.

Plant Protection: The papaya is susceptible for virus disease, which are spread through insect vector. The rouging of is strictly followed to avoid further spread of the disease, in addition the insecticidal sprays are taken against aphids, white flies and other sucking pests. Powdery mildews, downy mildew and stem rot. Providing wind breaks, well drained conditions to soil avoiding planting papaya after papaya and following a suitable crop rotation keeping the field clean and weedfree are the some of the measures which will help to keep the papaya crop in good healthy condition

Harvesting, Packing, Post Harvest Handling: Usually the fruits are harvested when they are full size, light green with a tinge of yellow at epical end. When the latex ceases to be milky and become watery the fruits are considered suitable for harvesting. First picking may start at 14/15 months after planting. Three to five picking for one season are often taken fetching about 30-35 tonnes per hectare. Suitable grading must be done before packing. Since the fruits are easily perishable, care should be taken to wrap the individual in paper and finally packed in crates.

Special Features—Papain Production: Crude papain in flakes or powder crude papain white brown. Spray dried crude papain are the three types of papain extracted from raw-mature papaya fruits. About 8-10 g of final produce can be obtained from a well developed fruit weighing about 2 kg. Tapping may be done 2/3 times preferably during cool hours. The fruits after latexing can be used as a table fruit or for preparing fruity or jams etc.

Kokam

Kokam (Garcinia indica chaisy) is an evergreen tree, native of India mostly found in Western Ghats along the seacoast in Konkan region of Maharashtra state.

Importance: The fleshy ripe purplish red fruit possesses sweetish to acidic pulp. Apart from traditional use for preparing "Amsul" or "Kokam Sol" and "Amrit Kokam" squash, it is gaining importance for its taste. The Kokam seeds contain 23-26 per cent fat which remains solid at room temperature. Hence, it is being used as a substitute for Coca butter. It is also being used in cosmetics. Due to this, demand for kokam both in internal market as well as in export market is increasing. It is, imperative to extend the area under Kokam plantation on a large scale using cultivable waste land in Western Ghats.

Climate: Kokam is a Tropical fruit ,grows and fruits well in high rainfall areas of the sea coast of about 1000 to 2000 mm having warm and humid climate.

Soil: Lateritic, alluvial or medium deep well drained soils are suitable. It is mostly grown in hill slopes mixed with other fruit trees like jamun, mango, cashew etc. in Kokan region of Maharashtra.

Varieties: There are no standard varieties. Kokam is a slow growing plant and exhibits a large variation in sex forms and bearing capacity. The Kokan Krishi Vidyapeeth has spotted a high yielding (35 kg seed/ tree) under natural conditions.

Such selected trees could be clonally propagated by soft wood grafting.

Planting Season: Pits of 60X 60 X 60 cm be dug during summer at 6 X 6 m distance and filled with well decomposed farm yard manure and top soil at a ratio of 1:3 and 1 kg superphosphate be added at the bottom of the pit. The planting of sapling be done at the onset of monsoon in June. At the time of planting 100g carbaryl dust (10 per cent) be mixed in each pit to avoid termite attack.

Interculturing: Field be kept weed free by regular weeding and hoeing. Short duration, low growing, inter crops like cowpea, vegetable can be grown during first four five years .

Care of Young Orchards: Young plants be given support of bamboo sticks. Shoots arising from the main stem just before the bud /graft joint be removed regularly. Wherever possible young saplings be irrigated during dry period and summer months for initial 2/3 years.

Special Horticultural Practices: Since it is grown as a rainfed crop no special treatment is required. However, if the orchard is established by planting seedlings about 50 per cent plants will be male which will not bear any fruits. This could be identified 7-8 years after planting when they start flowering. This could be converted by grafting with selected scion stick taken from female (fruit bearing) plants. However, it is necessary to maintain 10 per cent male plants in an orchard for pollination.

Irrigation: Kokam is grown as rainfed fruit crop and irrigation is not given except during summer or dry spells in the initial 2/3 years.

Nutrition: One year plant be given 2 kg FYM + 50g nitrogen + 25g P_2O_5 + 25 g K_2O during August- September. The quantity be increased by 2 kg FYM + 50 g nitrogen + 25g each of P_2O_5 and K_2O every year. From 10 years onwards each tree be supplied with 20 kg FYM + 500g N and 250g each of P_2O_5 and K_2O during August September.

Plant Protection: No serious pests/ diseases are reported on Kokam and hence it is normally does not require any plant protection measures.

Harvesting and Yielding: Seedling trees starts fruiting after 7-8 years while grafted/ budded plants bear fruits after 4-5 years. Flowering starts in October November and continues upto Feb-March. Fruits are ready for harvest during April-May. Red ripe fruits be harvested with the help of 'Atul' harvestor specially developed for this purpose by Konkan Krishi Vidyapeeth, Dapoli, Most of the existing plantations are by seed. Hence the yield varies from place to place depending upon orchard management practices. On an average individual trees yield 30-50 kg ripe fruits per year.

Post Harvest Handling and Marketing: Kokam fruits are juicy and highly perishable and hence need careful handling and after harvesting. Fruits are collected in bamboo baskets lined with rice straw and stored under shade. Harvested fruits are sorted removing under signed ,damaged fruits and graded as per the size and colour into two grades. Mostly fruits are processed into 'Amsul' or 'Amri Kokam' squash because of poor transportability and highly perishable nature.

Limes and Lemons

Lime or Acid lime (Citrus aurantifolia swing): Is more popular in India than lemon (Citrus limon). Acid lime is cultivated largely in Andhra Pradesh, Maharashtra, Tamil Nadu, Gujarat, Rajasthan, Bihar and to a limited extent in other states.

Lemons are grown only to a limited extent in homestead gardens. Limes are available throughout the year in some part or the other in our country. They are used as fresh fruit or for the preparation of pickles and beverages. They are rich in vitamin C, minerals and salts.

Climate: Areas with dry climate and low rainfall are best suited for growing limes. Lemons can be grown in heavy rainfall humid regions.

Soil: Lime and Lemon may be planted in medium black, loamy or alluvial soils having perfect drainage and devoid of calcium carbonate layer.

Varieties: Commonly grown acid lime cultivar is Kargzi lime, but recently identified four high yielding cultivars Vikram,

Prumalini, PKM, and Sai Sharbati are now gaining popularity. Commonly grown lemon varieties are Assam lemon, Italian lemon, Pant lemon, Galgal and Eureka lemon, Sevilla and Malta lemon varieties are popular in South India.

Propagation: Acid lime can be propagated by seed, budding or air layer. Since it is highly polyembryonic, seedlings are true to type and resistant to diseases. Propagation by seedling is recommended. Budded plants on Jambheri or Gajanimma are used for planting lemons.

Planting: Pits of box 60 x 60 cm size are dug 5 to 6 meters apart each way and filled with 30 to 40 kg. Farm yard Manure or compost and Top soil along with 2 kg superphosphate. One year old, healthy seedlings of recommend acid lime variety of bud grafts of choice lemon variety is used for planting. Most ideal season for planting is the onset of monsoon (June - July) for both time and lemon. However, in heavy rainfall area September-October would be appropriate planting season.

Interculture: Low growing vegetables, leguminous or green manure crops may be grown as inter crops between young trees upto 4 to 5 years of their age. Area around plants may be kept weed free by weeding and hoeing regularly.

Care of young Orchids: Soon after planting plants be supported with bamboo stick and irrigated thoroughly. Main stem of the young sapling be topped at a height of about one meter from ground level within 25-30 cm from cut end 3-4 well spaced branches be allowed to grow in different directions to form main limbs of the tree. Plants be protected against pests and diseases by spraying with pesticides and fungicides regularly.

Special Horticultural Practices: Acid lime flowers and produces fruits almost all the year round, but there are three major flushes in June-July, September-October and January-February. These are known as Mrig Bahar, Hast Bahar and Ambia Bahar respectively. Fruits of September-October (Hast Bahar) flowering harvested during summer months are most remunerative. To achieve this, irrigation be suspended for 1 to 11/2 months before actual flowering. During this period field may be cultivated and applied with manure and fertiliser towards the end and irrigation be resumed. Plants be sprayed with 1000 ppm cycocil twice during September-October which has been found to increase the fruit yield two times than the normal yield.

Nutrition: For each completed year of growth till 5 years, 20 kg Farm Yard Manure, 100 g urea and 1 kg Superphosphate be applied per plant in two equal split doses in June and December every year. Fifth year onwards application of 40 kg FYM, 400 g N + 200 g P + 400 g K per plant in two split doses in June-July and September-October be applied. In addition, 150 g N be applied per plant 15 days after fruit set. Manures and fertilizers are applied in a circular trench 20-30 cm deep about a meter wide around the periphery of the tree. Manures are thoroughly mixed with soil, and the tree irrigated profusely.

Irrigation: Young plantation needs regular watering during post-monsoon period at an interval of 10-15 days and at 8-10 days during summer months. Water is applied around the periphery of the tree in a ring and basin system so that tree trunk should not come in direct contact with water. The irrigation is suspended for 1 to 11/2 months during Bahar treatment. It is resumed during flowering. Bearing trees should get regular watering at 10-15 days during summer and 15-20 days interval watering at 10-15 days during summer and 15-20 days interval during winter.

Plant Protection: The major pests affecting the Acid lime and lemon plants include Lemon butterfly, leaf eating caterpillars, whitefly, leaf miner and mealy bugs. These pests can be controlled by spraying with Monocrophos, Endosulphon or Phosphamidon 1.5 to 2 ml.per litre of water at an interval of 10-15 days. Instead of repeating the same pesticide different pesticide be used at each spray.

Among the diseases Gummosis and canker are very devastating. For control of Gummosis, use ring and basin system of irrigation to avoid contact of water with main stem so that the disease causing soil borne fungus will not attack the stem. If disease is seen remove the affected portion of bark and 1 cm of surrounding healthy bark with knife and apply Bordeaux paste (one part copper sulphate plus one part quick lime and sufficient water to make paste) thoroughly on the wound.

Canker is caused due to bacterial (xanthomonas sp.) infection, which develops considerably during rainy season. It appears as a minute yellowish-brown spots on stem, leaves and fruits which turn into corky brown specks. Affected leaves turn yellow and drop down. Remove all the affected parts and fallen leaves and burn them. Spray the trees with Bordeaux mixture 3:3:50 by using 1.5 kg copper sulphate plus 1.5 kg quick lime in 225 litres of water or 1250 gm of copper oxychloride in 500 litres of water also can be sprayed. Repeat, if necessary at an interval of 10-15 days.

Harvesting and Yield: Maturity of citrus fruits depends upon the climatic conditions, nutrition, heat units and moisture availability. The period of maturity is shorter in acid lime (5-6 months) and longest in mandarin and sweet orange. The lime and lemon fruits are harvested when they just start changing. They are harvested in installments as they ripen. Major harvesting period in July-September and November-January.

Post Harvest Handling and Marketing: Harvested fruits are graded according to size and colour and packed in bamboo baskets or wooden crates lined with neem foliage. Sometimes fruits are packed in gunny bags and transported by rail or road to distant markets.

Banana

Banana is the oldest and commonest fruit known to the mankind. It is one of the important fruits, and constitutes second largest fruit industry in India. It is nutritious palatable and easily digestible fruit. It is available throughout the year. Banana is rich in carbohydrates, minerals such as calcium, potassium, Mg, Na and phosphorous. Other than fresh fruits, it can be consumed as processed in various forms like chips, powder, flakes, etc. Banana pseudostem is chopped and used as cattle feed. Also, the leaves are used as plate. The botanical names of banana are Musa cavendish and Musa paradisiaca, which belongs to the family Musaceae.

Climate: Being a tropical crop, banana requires warm, humid and rainy climate. The optimum temperature range is 10 to 400C and the relative humidity is 90% or above. It is highly susceptible to frost and cannot tolerate arid conditions. Strong desiccating winds cause considerable reduction in the growth of the plant and yield and quality of fruits.

Soil: Banana is a heavy feeder crop/. Therefore, fertility of soil is very important. Rich, well drained, fertile, free working, soils with plenty of organic matter are best suited for cultivation. The optimum range of pH of soil should be 6 to 8.

Propagation: Commercial edible bananas do not produce viable seeds. So, the banana is commonly propagated by suckers and sword suckers with narrow leaves. Rhizomes whole or in bits from fruited and non-fruited plants with atleast one sound bud can be successfully used as propagating material.

Planting: Planting of banana is done by two methods viz. Pit method and furrow method. Planting is done from February to May whereas in North India, it is done during July-August. In South-India, it can be done any time of year except summer. Tall varieties should be planted at 3x3 m whereas dwarf ones at 2x2 m apart.

Manuaring: Banana is a heavy feeder and responds well to manuring. Banana is a quick growing and short-lived plant. Therefore, it is more beneficial, if quick growing fertilizers are applied. In Tamil Nadu, application of 100 g of N, 30 g of P_2O_5 and 300 g of K_2O per plant to be supplied in three doses during the second, third and fifth months is recommended. In Maharashtra, 100 g N, 40 g P_2O_5 and 100 g K_2O per plant is recommended. Out of these, P_2O_5 and K_2O is applied at the time of planting and N is given is three split doses: third, fourth and fifth month after planting.

After Cares:

i. *Desuckering:* Desuckering is an important operation in banana cultivation. In this, the unwanted suckers, which develop near the base of the plant, are removed. Cutting the suckers from ground level and then pouring the kerosene (2-3 drops) to kill the growing point of pseudostem does it.

ii. *Propping:* It is an essential cultural practice to give proper support to the plants with bamboos. It is done to avoid falling down of plants due to high winds.

iii. *Wrapping:* To protect the fruits from sunburn, hot wind and dust the bunch is covered. Wrapping is also done to improve the colour of the fruit.

Harvesting and Yield: Harvesting of banana is done 12 to 15 months after planting in dwarf and 15 to 18 months after planting in tall varieties. Signs of maturity of banana fruits are, fruit becomes plumpy and angles are filled in completely, when tapped gives metallic sound, drying off of top leaves and change in colour of fruits from deep green to light green.

Tall varieties like Poovan yield 15-25 tonnes/ha, while Dwarf Cavenshish yield 25-50 tonnes/ha. It can be stored at temperature slightly above 550F and relative humidity of about 85-95% for about three weeks.

Varieties: Following are the major banana varieties grown in India.

A. *Poovan:* It is the most important commercial variety in Tamil Nadu, Andhra Pradesh and West Bengal. It is also known as Lal velchi in Maharashtra. It is resistant to Panama wilt,

B. *Dwarf Cavendish or Basarai:* It is a dwarf variety. It is resistant to Panama disease. It is a high yielding variety with fruits large and of good quality.

C. *Robusta or Harisal:* Fruit colour remains green when ripe. It is best variety for the export purpose.

D. *Rasthali or Mutheli:* It is a good variety but susceptible to Panama disease

E. Rajeli or Nendran

F. Sonakela

G. Safed Velchi.

Plant Protection:

Disease: Important diseases of banana are:

Panama Wilt: It is caused by the soil bourne fungi Fusarium oxysporum.

Symptoms: Leaves become yellow, Pseudostem splits and breaks.

Control Measures: Eradication of the infected plants. Growing resistant varieties like Dwarf Cavendish. Soil drenching with Vapam @ 0.85% and mercuric chloride @ 3000 PPM in nursery.

Bunchy Top: It is caused by the virus, which is transmitted by an aphid Pentalonia nigroneruosa. Symptoms: Infected plants have short, narrow, erect leaves with short petiole. Plants remain slanted.

Control Measures: Infected plants should be uprooted and burnt. Spraying of 2-4, D to control weeds.

Insect Pests: Important insect pests are:

Root Stock Weevil: Damaging stage is adult and grub

Symptoms of Damage: Grub bores into stem which is affected by fungi or bacteria.

Control Measures: Use healthy suckers and rhizomes. Before planting treat the pit with 0.65 %. Lindane 60-80 g/pit. Soak the sucker in 0.1% Lindane solution, spray 0.05% endosulfan.

Pineapple

Pineapple is Tropical Fruit: It is a good source of Vitamin A and B and rich in Vitamin C and calcium. It contains phosphorus and iron. The fruit contains a special enzyme called 'Bromelin' which digests protein. Fruit is used for table purpose, preparations of juices, slices, tit bits, squash, jam, mixed jam, candy etc. Dried waste after extraction (pomace) is valuable cattle feed. Other byproducts are alcohol, calcium citrate, citric acid and vinegar. Leaves of pineapple are used for fibre preparation.

Fruit of pineapple is botanically called sorosis. Botanical name of pineapple is Ananas comosus and family is Bromeliaceae.

Climate: Pineapple is tropical fruit. It requires moderate temperature and high humidity. The optimum temperate range is 210C to 230C. At low temperature, no fruit bud differentiation takes place. A well-distributed rainfall of 150 c. m. per year is adequate.

Soil: Crop is grown on light to medium soils. Soils should be well drained. Sandy loams, laterites and slightly acidic soils with pH 5.5 to 6.00 are suitable for crop cultivation.

Propagation: Pineapple is propagated by vegetative method, suckers, slips, crown and disc are used for its multiplication. Slips are best for planting.

Planting: For planting rainy season is the best, planting is done at beginning of monsoon in dry region and at the end of the monsoon in high rainfall area. Planting is done either in flat beds or in shallow trenches. The planting distance in the single row system should be 25 x 60 x 90 cm and in double row system 25 x 30 x 90cm.

Manuring: Pineapple is heavy feeder. It requires 600 kg N, 400 kg P and 400kg K per ha. Nitrogen is given in two split doses, first at begining of monsoon and second in the month of February.

Harvesting and Yield: It matures in both 15 to 20 months after planting. Usually flowering takes place from February to April and fruits are ready from July to September. The fruits are harvested when they just become yellow. An average yield is 10 to15 tonnes per hectare.

Varieties:

i. *Gaint Kew or Kew:* It is one of the most important cultivars of Pineapple. It is ideal for canning industry. Flesh of the fruit is yellow, fibreless and juicy. The fruit is cylindrical in shape with average weight 1.9 kg. It is a late maturing variety.

ii. *Queen* : It is the second important cultivar which is grown in our country. It is an early maturing variety, used as table variety. It is mostly grown in Bengal region. The fruits are small cylindrical, with average weight 1.2 kg. Fruit colour is golden yellow while the flesh is deep golden yellow

iii. *Mauritius* : It is a mid season variety. Fruits of this variety are medium sized with yellow and red skin.

iv. Typhone No.1 and No.3

v. Red spanish

Fruit Culture

Fruit Culture: India has a wide variety of climate and soil on which a wide range of Tropical, Sub-tropical, Temperature and arid zone fruit crops can be grown in different regions.

The important advantages of fruit culture are:

i. From an unit of land comparatively more income is realised by growing fruits than growing any agronomic crop like wheat, rice, maize, etc.

ii. Calorific value of food from an area of fruit is more than cereals *e.g.* banana is 16 times more than wheat.

iii. Fruit farming is the bad rock of several industries like extracting of essential oil, pectin, production of spices, fruit preservation industries, etc.

iv. Fruit culture provides employment to the rural people throughout the year.

v. Cultivation of fruit crops helps in maintaining ecological balance and checks the pollution.

vi. Being rich in minerals and vitamins fruit crops assume great importance as nutritional security of the masses.

Based on the temperature requirements and response to different climatic conditions, fruit crops have been classified into three broad groups.

Tropical Fruits: Are those which do not endure severe cold but can tolerate warm temperatures of about 40 deg. Cent. These plants need strong sunshine, moderate warmth, fair degree of humidity and a very mild winder. They cannot stand frost or snow. Papaya, banana, pineapple, sapota and cashew belong to this group.

Sub-tropical Fruits : Like oranges, guava, mango, litchi, pomegranate, fig and Amla are intermediate in character. They need warmth and moderate humidity and can also tolerate mild winters. These plants are versatile in their ability to stand extremes of rainfall, humidity, sunshine and mildly tolerant of winter upto frost stage.

Temperate Zone Fruits: Are commonly found in cold regions enjoying a mild and temperature climate. These fruit trees endure cold and go to rest or dormancy by shedding off all their leaves during winter, apple, pear, peach, plum, apricot, Cherry, and grapes are examples of such fruits.

The above classification indicates the broad differences in climatic needs of various fruit plants. This does not necessarily mean that a plant belonging to one climatic zone does not grow in other zones. For instance, grape vine is a fruit which goes to dormancy in temperate regions and remains evergreen in sub-tropical regions. It is grown in both these climatic zones as a result of accumilization, cultural manupulations and breeding new varieties.

Mango

Mango (Mangifera Indica) is the most ancient among the tropical fruits and believed to have originated in the Indo-Burma region. India is the major mango producing country in the world with an annual production of 8.50 million tonnes from an area of one million hectares.

Mango is basically a tropical plant but endures wide range of temperature. It grows well under tropical and sub-tropical conditions. It gives profitable yield in semi-arid conditions, especially with irrigation. It grows where temperature drops as low as 0 deg. Cent. And as high as 46 deg.cent. However, it thrives best at temperature around 27 deg. Cent.

It grows well both in low (25 cm) and high (250 cm) rainfall areas. However with annual rainfall around 75 cm, it grows withoug irrigation. The mango flowers and fruits during dry season, which is chracterised by absence of rainfall. Rain or cloudy weather at the time of flowering causes considerable damage to mango as it adversely affects flowering and fruit set and increases incidence of pests and diseases.

Mango needs a deep, fertile and perfectly drained soil. Alluvial, lateritic and medium black soils are considered good. Soil depth should be atleast 2.5 m. and soil pH between 5.5-7.5 is good.

Varieties: The important commercial mango varieties for different parts of India are:

North India: Dashehari, Langra, Chausa, Bombay, Green, Somar Bahist

Eastern India: Fazli, Krishna Bhog, Himsagar, Gulab Khas, Jardalu

Western India; Alphonso, Kesar, Pairi, Mankurd, Fernandin.

South India: Bangalore (Totapuri), Neelam, Baneshan, Badami, Suverna rekha.

Newly Evolved Varieties: Mallika, Amrapali, Arka Aruna (Hy 10), Arka Puneet (Hy 13), Arka Anmol, Ratna, Sindhu, Sai Sugandh.

Orchard establishment and care:

Pits of 1 x 1 x 1 m are dug at a distance of 10 x 10 meter at plant site in summer. Pits are filled with top soil and 5 to 6 baskets of Farm Yard Manure (FYM) and 2.5 Kg. Super Phosphate. To avoid termite attack 100 gm of 10 percent sevin or chlorudane dust is mixed in the pit.

Planting of mango grafts of desired variety is done with commencement of rains while in heavy rainfall regions. Planting is best done after censation of heavy rains. Selected healthy grafts with smooth union and true to type, obtained from reliable source should be planted. At the time of planting, the graft is carefully placed in the pit after removing the Polythene bag and soil is packed firmly around the roots. It is irrigated immediately and stalked with bamboo sticks. Young plants are trained by removing lower branches upto half meter from ground level, only 4 to 5 well spaced branches are allowed to grow in different directions to form main limbs of the Tree. Shoots arising on stem below graft joint are removed regularly.

In young orchards, vegetables and other low growing crops like groundnut, blackgram, green gram, cowpea are grown as intercrops/ covercrops. Cultivation helps to check the weed and improves the growth of the trees besides giving some income to the growers till mango plants start bearing 4-5 years after planting.

A young orchard be supplied with 10 kg FYM + 100 Kg N + 50 g P_2O_5 + 100 g K_2O per tree in the first year and it should be increased with age. The 10 year old tree should get FYM + 1 Kg N + 500 g P_2O_5 + 1000 g K_2O. It irrigation is available, additional dose of 500 g N be given in the month of February-March.

Newly planted grafts be irrigated at 3-4 days interval for 6 months, thereafter interval should be 8 - 10 days depending upon climate Irrigation interval of 10-15 days is adequate for 1-5 year old plants. For bearing trees, irrigation should not be given prior to flowering for 2-3 months as it promotes vegetative growth and reduce the crop. When the fruits have developed to the size of a marble, irrigation can be started and continued at weekly or fortnightly intervals. This helps to reduce fruit drop, promotes rapid fruit development, early maturity and improves quality. However, mango is mostly grown as rainfed crop.

Harvesting: Mango fruits mature in 3-4 months from flowering, Fruit colour changes from dark green to light green on maturity. Harvesting should be started after few fruits drop, during morning hours. Individual fruits are clipped with 1.5 cm stalk using mango harvester. Harvested fruits are kept on gunny bags under shade. Under sized, bruized and infested fruits are sorted out and healthy fruits are graded into 2-3 grades depending upon colour, shape and size and packed in wooden or corrugated boxes.

Export Potential: India is the largest producer of mangoes in the world accounting for over 63 percent of total world production. It is the most important fruit grown in the country, which accounts for over 60 percent of the total value of fruits exported from the country. A substantial quantity of this fruit is also utilised by the fruit processing industry.

Maharashtra has clear advantage due to exclusive production of Alphonso variety, the share of which in the export is very high.

Plant Protection: Mango hoppers insectpest and powdery mildew disease are the major problems on mango. To control both these dusting with 10 percent Carbaryl dust plus Sulphr

300 mesh in the ratio of 1:1. Three to four times at fortnightly intervals starting from prior to the emergence of first inflorence is recommended. Trunks of the tree should also be dusted thoroughly.

Ber

Ber (Zizyphus mauritiana Lamk) is an ancient and indigenous fruit of India, China and Malaysia region. The fruits are very nutritious and are rich in vitamin C, A & B complex. The ber is one of the most common fruit trees of India and is cultivated practically all over the country. Ber fruits can be within the reach of the poor people and hence known as poor man's fruit.

Importance: Among the fruit trees, ber cultivation requires perhaps the least inputs and care. It gives good production even without irrigation and can be grown as a rainfed crop in semi-arid and arid regions. The tree can, therefore give assured income even under marginal growing conditions and provides nutritious food at very low cost. The fruit is dried and is used as a dessert fruit. It can also be preserved as a candied fruit.

Climate: The ber is a hardy fruit and grows well all over the country under varying climatic conditions and upto an elevation of 1000 meters above sea level. For its successful cultivation, it favours a hot and dry climate. The Ber crop withstands high temperature and aridity by cessation of growth, leaf fall and dormancy phase. In high humidity conditions disease and pest problems increase especially of powdery mildew which is a serious disease.

Soil: Ber plant grows on a wide variety of soils ranging from shallow to deep and from gravelly and sandy to clayey. The ber develops a deep tap root system within a short period of its growth, and as such to adverse soil conditions. Ber can also withstand alkalinity and slightly waterlogged conditions. It can, in fact, withstand and often do better than most fruits on poor soils.

Varieties: Numerous horticultural varieties of ber are grown all over India. Some of the most popular varieties are Umran, Karaka, Gola, Seb, Chhuhara, Sanaur - 2, Ilaichi and

Mehrun. In Maharashtra, cv.Umran Is being grown on commercial scale. It has excellent keeping quality and transportability but flat taste.

Propagation: The Ber was commonly propagated by seeds during earlier period. But main disadvantage of this method is of heterozygosity and variability in seedling progeny. Therefore, propagation of superior varieties by patch budding is recommended. For raising a budded plantation, it is considered best to sow ber seeds in the field itself at proper distances and use the seedlings thus raised for budding in situ. For early germination of ber seeds, breaking of endocarp (hard seed coat) gives quick germination.

Planting Season: Monsoon season offers best choice, for raising in situ plantation in Maharashtra and arid and semi-arid parts of our country, at a spacing of 6 x 6 m. Protective irrigation during summer months will ensure the survival and good growth of the plants. Deheading these seedlings to ground level in the month of May helps to give new shoots by July. Patch budding these shoots in July helps to convert it into choice variety. In Northern India, planting is done either in February-March or July-September at a spacing of 7-8 months.

After the layout, pits of 60x60x60 cm are dug. About 100 g of 10 per cent carbaryl or Aldrex dust is sprinkled on the bottom sides of pits to prevent termites. Pits are filled with top soil mixed with 20 kg farmyard manure and 1 kg super phosphates. Treated seeds or bud grafts are planted in these pits at the onset of Monsoon.

Interculturing: Area around the young plant is kept clean by weeding and hoeing. Under irrigated conditions low growing vegetables can be grown as inter crops until the ber plants assume (4-5 years) full grown. Under rainfed conditions legume crops like Moong, Moth (horse gram) and cowpea can be grown as intercrop.

Stirring the soil under the tree canopy after rains provides better aeration around roots and helps in conserving soil moisture and weed control.

Care of young orchards:

Training: If left to itself ber trees would attain bushy, large unmanageable form but the production per unit area is quite low. In order to keep the plant in manageable shape and size, trees be trained properly during first 2-3 years to build a strong framework. By providing support with bamboo stick to the new growth of sprout from either in situ or transplanted seedling, vertical growth is encouraged Once the growth is about 1-1.5 meter, the terminal growth Is pinched, allowing lower buds on main stem to sprout and form main branches. After sufficient growth, these primary branches be pinched to develop secondary branches. Thus plants are trained by pruning and kept in manageable shape, size with well developed framework.

Pruning: Pruning is an essential operation in ber production as fruits are borne in the axil of leaves on the young shoots of current season. Pruning is therefore, done every year to induce maximum number of new healthy shoots which bear good quality fruits. The best time for pruning was observed to be 15th April to 15th May under prevailing climatic condition of Maharashtra.

Special Horticultural Practices: Attempts to increase fruit set, fruit size and early/delayed maturity through application of growth regulators proved to effective.

The growth regulators like GA, NAA, CCC and Ethephon were used. GA (10 ppm) and NAA (10 ppm) during fruit development and Ethephon one month before harvesting were sprayed. Application of NAA increased the yield of fruits with cost benefit ratio of 1:3:30.

Irrigation: Ber is mostly grown as a rainfed crop. Irrigation as a rule not advisable. But if available could increase yield. However, increase in yield also diminishes the quality of fruit.

Nutrition: Like other fruit trees, ber also requires regular application of manures and fertilisers for good yields. Application of 30-50 kg Farm yard manure, 250 g N per tree (split in 2 equal doses) and 250 g P_2O_5 per tree + 50 g K_2O per tree (single dose) per year for full grown tree (5 years and thereafter) is recommend for better yield and quality of fruits.

Plant Protection: Ber fruitfly is one of the important pests of ber, which is widely distributed throughout India. The infested fruits turn brown, rot and smell offensively. The pest can be controlled by spraying 3 to 4 times with Carbaryl 50 WP 0.2 per cent or Dimethoate 30 EC 0.03 per cent commencing from the attainment of fruits of the pea size.

Among the diseases powdery mildew is very common on ber fruits. Small whitish spots appear on young fruits, which later enlarge and cover the entire fruit. The affected fruits either drop off or become corky, mis-shapen and under-developed. Disease can be controlled by dusting with sulphur @ 150 to 200 g /tree and subsequent 3 dustings at an interval of 15-20 days.

Alternaria leaf spot and Cercospora leaf spot also appears in the form of grey spots on leaves. Both these diseases are effectively controlled by spraying dithane M-45 (0.25%) or Foltart (0.1%) as soon as the disease appear. Subsequent 2 to 3 sprays be given at an interval of 15 to 20 days depending upon the intensity of the disease.

Harvesting and yield: In Northern India, peak period of harvesting falls between February and April. While in Maharashtra harvesting extends from November to January. The fruits are harvested in 4 or 5 pickings since all the fruits on the tree do not mature at one time. The fruit picking is done by hand using a ladder. The fruits should be harvested at proper stage of maturity. The best index of the correct picking stage is the characteristic maturity colour and softness of particular cultivar after the fruit has attained the full size. The fruit requires about 120 days to reach maturity.

Under dryland (rainfall) conditions, on an average 60-80 kg fruits per tree per year can be harvested. Under irrigated conditions yield will be 3-4 times higher.

Post harvest handling and marketing: The underripe, overripe and damaged fruits are sorted out. The remaining sound fruits are graded in two grades -large and small according to size. Fruits are packed in gunny bags, wooden boxes, cardboard boxes or nylon knot bags.

The growers generally auction their crop to the contractors. The contracts are fixed either on the share of accruing incoming basis or on the basis of a lump sum to be paid in installments to the grower.

Grapes

Introduction : The grape is the most important crop grown in the world. Mostly it grown for making wines and preparation of raisin and then as a table fresh fruit. While in India, it is mainly grown for table use. Grape cultivation is believed to have originated near Caspian Sea, however, Indians know grapes since Roman times. Total area under grapes in India is about 40,000 ha, distributed mainly in Maharashtra, Karnataka, Andhra Pradesh and Tamil Nadu.

Economic Importance: At present, grape is the most important fruit crop grown commercially with the objectives.

a. For table purpose
b. For export purpose
c. For making wines and
d. For making raisins.

Fresh grapes are a fairly good source of minerals like calcium, phosphorous, iron and vitamins like B. Famous champagne and other desert wines are prepared from grapes.

Climatic Requirements: The ideal climate for grape growing is the Mediterranean climate. In its natural habitat, the vines grow and produce during the hot and dry period. Under South Indian conditions - vines produce vegetative growth during the period from April to September and then fruiting period from October to March. Temperatures above 100C to 400C influence the yield and quality. High humidity and cloudy weather invite many fungal diseases, besides lowering the T.S.S. : Acid ratio.

Soil Preference: The grape is widely adopted to various soil conditions, but the yield and quality reach to the highest on good fertile soils have pH 6.5 to 8.5, organic carbon above 1.0%, free of lime and having a medium water holding capacity. Early but medium yields with high T.S.S. are harvested on medium type of soils.

Varieties: Table purpose varieties:

i. *Seeded Varieties:* Cardinal, concord Emperor, Italia, Anab-e-shahi, Cheema sahebi, Kalisahebi, Rao Sahebi,

ii. *Seedless Varieties:* Thompson seedless, flame seedless, kishmish chorni, perlette, Arkavati.

Raisin Purpose Varieties: Thompson seedless, manik chaman, sonaka, Black corinth, Black monukka, Arkavati, Dattier

Wine Varieties: Chardonnay, Cabernet Saurignnon, Bangalore Blue, Muscat, Blanc, Pinot Noir, Pinot Blane, White Riesling, and Merlot.

Propagation: Grapevine is most commonly propagated by hard-wood cuttings, though propagation by seed soft wood cutting, layering, grafting and budding is specific to certain situations. Occasionally, unrooted cuttings are also planted directly in the field in the pre-determined position for a vine. For hardwood cuttings, IBA, 1000 ppm treatment is useful for early, better and uniform rooting of cutting. For grafting Dog ridge, Ramsey, 1616, 1613,1103P, So4, etc. are used. Sometimes the rootstocks are planted in the field and there they are grafted with suitable varieties.

Planting and Season:Usually planting is done from October onwards till January. Rarely planting is also done during June-July where the monsoon is late. Monsoon planting is avoided mainly for avoiding diseases on young growth. For planting N-S direction the trenches are opened. The size of trench may be 60 to 75 cm. Deep wide. Then these trenches are filled with FYM, organic manures, 5:10:5 organic mixtures, single super phosphates, biofertilisers, neem cakes, etc. Spacing for planting is maintained depending on soil type, variety and method of training. The distance between two rows may be 2 to 3 m while distance between vines within a row will be half of that, accommodating vines from 2000 to 5000 per hectare.

Interculturing: The following aspects are important:

i. *Gap Filling:* To be done preferably during one month after planting.

ii. *Recut:* Basal cut keeping 2/3 buds is taken one month after planting with an objective to get uniform new growth.
iii. *Supporting:* The bamboo supports are fixed for vine support and young growing points are trained on them
iv. *Weeding:* The vine rows are weeded out, twice/thrice depending on the intensity of weeds.
v. *Irrigations:* regular depending on soil and season are given.
vi. Fertilizers are applied with cow during slurry to hasten the growth.
vii. Suitable plant protection measures are following depending on incidence of pests and diseases.

Grapes vines takes about 1.5 to 2 years after planting to bear the first crop. During this period the care of young vines is taken as under:-

i. *Training:* The vines are trained first on bamboo and then on support—trellis. A suitable method of training is adopted.
ii. *Pruning:* Initial pruning is done only for training *i.e.* for developing trunk, arm, fruiting, canes, etc.
iii. *The Fertilizer Doses:* Including organic, inorganic and bio-fertilizers are applied twice in a year.
iv. Plant protection schedule is prepared and followed for the total initial period of growing.

Special Horticultural practices:

i. *Pruning and Training:* The vines are trained on a suitable trellis *i.e.* 'T', 'Y', 'H' or bower and regularly pruned twice in a year. First annual pruning is done during the month of April to get the new vegetative growth while second pruning to get the crop is done during the month of October. While doing April pruning 0 to 2 buds on arm are kept while doing October pruning 5 to 10 buds on fruiting cane are kept. Use of HCN is done to have early, uniform and higher sprouting particularly after winter pruning is made.
ii. *Girdling:* Vines are trunk girdled at bloom period to

increase the fruit set, to increase the weight and T.S.S. and also to enhance maturity.

iii. *Use of Hormones:* The following plant hormones at various stages and concentrations are usually used to increase the yield and to improve the quality of bunches.

A.	At bloom	GA3	20 to 30 ppm; CCC	500 ppm
B.	At setting	GA3	30 to 40 ppm; 6BA	5-10 ppm
C.	At 4-6 mm size	GA3	40 to 50 ppm; Brassinos	100 ppm
D.	At 6-8 mm size	GA3	30 to 50 ppm; Cppu	2-3 ppm

Irrigation: Grape is strictly irrigated perennial crop and regularly irrigated. For flood irrigation, 5-7 days during summer 8-10 days during winter and 15-20 days during rainy season-interval is maintained while for drip irrigation, 40-50 L; 30-40, 20-30 L of water per day per vine, water is applied.

Nutrition: Balanced nutrition and use of chemical, organic and bio-fertilisers is essential to get a good crop of good quality every year. About 700 to 900 N, 400 to 600 P and 750 to 1000 K Kgs/ ha/year are applied to get about 30 to 35 tonnes produced yearly.

The use of vermiphos, biomeal, mixtures of 5:10:5 ormichem, micronutrient mixtures have proved useful in grape production. Fertilizers are applied mainly twice in a year at the time of pruning, besides occasional foliar sprays are also practiced. Now-a days, Fertigation techniques is being popular in grape growers.

Plant Protection: Grape shoots, leaves, blossoms and berries are attacked by many fungi and insect pests, besides some nematodes are also cause damage to roots.

Major Fungal Diseases: Anthracnose, Powdery mildew, Downy mildew,

Dead arm, Botraytis and Botrodiplodia.

Major Bacterial Diseases: Xyantomonas, blight.

Viral Diseases: Fan leaf disease.

Major Insect Pests: Flea beetle, Mealy bug, Red mite, Thrips, caterpillars.

Soil Born Pests: Nematodes, phyloxere, white ants, and white grub.

Control Measures: A number of systemic and contact fungicides and pesticides are available and are to be used as per following local schedule.

Besides pests and diseases the crop is to be protected against weeds, Cyprus, doob grass, Parthenium olerace are some of the common and important weeds found in vineyards. They are controlled by frequent weeding fruit ching/growing cover crops or by using chemical weedicides as Gramaxone, Basaline, Roundup, Glycel, etc.

Grape bunches are also to be protected against hot sun, Cold wave, dry air spell, Dew and Storm. Some chemical some physiological and some mechanical means methods are adopted.

Harvesting and Yields: Normal grape harvest season starts in February and continuous up to end of April. Well matured bunches having at least 180 Brix are harvested

Av. yields - For seedless varieties - 20 to 30 t/ha/y

For seeded varieties - 40 to 50 t/ha/y

Post Harvest Handling: Harvested grapes are packed in 2 to 4 kg-corrugated boxes. Grape guards, pouches are kept inside the boxes for distant markets. Pre-cooling and use of grape guards are the musts for cold storage and export markets.

Mumbai, Delhi, Calcutta, Ahmedabad, Ludhiana, Patna, Jamshedpur, Bangalore, Hyderabad, are the main market places in the country.

Some additional features:

a. *Export:* Day by day increased quantities are exported to Europe, Middle East, Dubai, etc.
b. *Winemaking:* Wine making and champagne making are profitable and started at few places in Maharashtra and Karnataka.
c. *Raisin Making:* A good quality black and golden raisins are prepared from the varieties. The raisins are increasing demand from Indian as well as from outside countries.

Strawberry

Strawberry (Fragaria sp.) is a native of temperate regions, but varieties are available which can be cultivated in sub. Tropical climate. In India it is generally cultivated in the hills. Its main centres of cultivation are Nainital (district) and Dehradun in Uttar Pradesh, Mahabaleshwar (Maharashtra), Kashmir Valley, Bangalore and Kalimpong (West Bengal). In recent years, strawberry is being cultivated successfully in plains of Maharashtra around Pune, Nashik and Sangali towns. The strawberry is the most widely adapted of the small fruits. Strawberries are grown throughout Europe, in every state of the United States, as well as in Canada and South America. The wide variation in climates within these regions and the wide adaptation of the strawberry plant permit harvesting and marketing, the fruit during greater part of the year.

Importance: Strawberry is a delicious fruit taken fresh in several ways. It also makes excellent ice cream and Jam on account of its rich aroma, and is also a good source of vitamin C. It is a soft and a highly perishable fruit, often shipped in frozen condition in Western countries.

Climate: Strawberry thrives best in temperate climate. It is a short day plant which requires exposure to about 10 days of less than 8 hours sunshine for initiation of flowering. In winter, the plants do not make any growth and remain dormant. The exposure to low temperature during this period helps in breaking dormancy of the plant. In spring when the days become longer and the temperature rises. The plants resume growth and begin flowering. The varieties grown in milder sub. Tropical climate do not require chilling and continue to make some growth during winter.

From the standpoint of response to length of the light period, strawberries are placed in two groups: (1) varieties that develop flower buds during both long and short light periods, the everbearing varieties and (2) varieties that develop flower buds during the short light periods only, most commercial varieties.

Soil Requirement: Strawberry requires a well-drained medium loam soil, rich in organic matter. The soil should be slightly acidic with pH from 5.7 to 6.5. At higher pH root formation is poor. The presence of excessive calcium in the soil causes yellowing of the leaves. In light soils and in those rich in organic matter, runner formation is better. Strawberry should not be cultivated in the same land for a number of years. It is preferable to plant it in green manured field. Alkaline soils and soils infected with nematodes should be avoided.

Varieties: A large number of varieties are available. For the hilly areas, varieties Royal Sovereign, Srinagar and Dilpasand are suitable. Some of the introductions from California, such as Torrey, Toiga and Solana may prove even more successful. The variety found successful in Bangalore has been named Bangalore and which has performed well at Mahabaleshwar also. For the north Indian plains, Pusa Early Dwarf which has dwarf plants, large firm wedge-shaped fruits, has been recommended. Another variety with rich aroma but softer fruits is Katrain Sweet. Some of the varieties found successful in warmer parts of the U.S.A. are: Premier Florida-90, Missionary, Blackmore,Klonmore & Klondike. Some of these may prove successful for cultivation in Indian plains.

Propagation: Propagation is done by means of runners that are formed after the blooming season. The plants may be allowed to set as many runners as possible but not allowed to set any fruits. All the plants with good root system should be utilised to set a new plantation. Given the best attention and care, a single plant usually produces 12 to 18 runners.

Planting: The land for strawberry planting should be thoroughly prepared by deep ploughing followed by harrowing. Liberal quantities of organic manure should be incorporated in the soil before plating. Strawberry can be planted on flat beds, in the form of hill rows or matted rows, or it can be planted on raised beds. In irrigated areas, plantings on ridges is advised. In Mahabaleshwar, the usual practice is to plant on raised beds 4 x 3 meters or 4 x 4 meters.

The planting distance should be 45 cm from plant to plant and 60 to 75 cm. from row to row. In the hills, Transplanting is done in March-April, September-October, but in the plains, the months of January-February may be utilised for this purpose. At Mahabaleshwar normally strawberry is planted during November-December.

The plants should be set in the soil with their roots going straight-down. The soil around the plant should be firmly packed to exclude air. The growing point of the plant should be just above the soil surface. During planting, the plants should not be allowed to dry out and should be irrigated immediately after planting.

Care of Young Plantation: The roots of strawberry plants spread out close to the surface. Therefore, the soil should be well-supplied with moisture, and hoeing should be done lightly and young plantation be kept weed free.

Special Horticultural Practices: In cold climate the soil is covered with a mulch in winter to protect the roots from cold injury. The mulch keeps the fruits free from soil, reduces decay of fruits, conserves soil moisture, lowers soil temperature in hot weather, protects flowers from frost in mild climates and protects plants from freezing injury in cold climates. Several kinds of mulches are used, but the commonest one is straw mulch. The name strawberry has been derived from this fact. Black alkathine mulch is also used to cover the soil. It saves irrigation water, prevents the growth of weeds and keep the soil temperature high.

Irrigation: Since strawberry is relatively shallow-rooted, it is susceptible to conditions of drought. Planting early in autumn allows the plants to make good vegetative growth before the onset of winter. However, in this case it is necessary to ensure that newly planted runners are irrigated frequently after planting, otherwise the mortality of the plants becomes high.

During September and October, irrigation should be given twice a week if there is no rain. It may be reduced to weekly intervals during November. In December and January, irrigation may be given once every fortnight. When fruiting starts, the irrigation frequency may should again be increased. At this stage frequent irrigation gives larger fruits.

Nutrition: Strawberry requires moderate amounts of nitrogen. Addition of organic matter to the soil, in the form of 50 tons of Farm Yard manure per hectare is highly desirable. It improves the water holding capacity of the soil and also gives better runner formation. Farm yard manure may be supplemented by chemical fertilizers to make up the total quantity of nitrogen from 84 to 112 kg per hectare, Phosphorus 56 to 84 kg per hectare, and Potash 56 to 112 kg per hectare. The Phosphatic fertilizer should be incorporated into the soil before plantings. The nitrogenous fertilizer be applied in Two doses (Three weeks after planting and again at the time of flowering) and potash at the time of flowering only. Application of adequate amounts of nitrogen gives higher yield of early berries.

Plant Protection: Red spider mites and cutworms are important pests of strawberry. The mites can be controlled with 0.05 per cent Monocrotophos + 0.25 per cent wetable sulphur. The cut worms can be controlled by dusting the soil before planting with 5 per cent chloradane or Heptachlor dust at the rate of 50 kg per hectare and mixing it thoroughly in the soil by cultivator.

The two commonest diseases of strawberry are red stele, caused by the fungus Phytophthora fragariae and black root rot. The remedy for the former lies by growing resistant varieties like stelemaster and for the latter to maintain the vigour of the plants and rotate strawberry with other crops like legume vegetables (beans, peas etc.). Strawberry also suffers from virus diseases known as yellow edge, crinkle and dwarf. Raising of strawberry nursery in the hills helps to check these. Strawberry also throws some chlorotic plants, which result from genetic segregation. These should not be confused with virus affected plants and should be rogued out.

Harvesting and Yield: The fruit ripens during late February to April in the plains and during May and June at high elevations like Mahabaleshwar, Nainital and Kashmir. For local market the fruit should be harvested when fully ripe, but for transport to distant markets, it should be harvested when still firm and before colour has developed fully all over the fruit. Harvesting should be done preferably daily. Since fruit is highly perishable, it is packed in flat shallow containers of various types (card board, bamboo, paper trays etc.) with one or two layers of fruits. Harvesting should be done early in the morning in dry conditions. Washing the fruit bruises it and spoils its lustre.

The yield varies according to season and locality. A yield of 20 to 25 tons per hectare is excellent, though yields upto 50 tons per hectare have been reported under ideal conditions.

Post Harvest Handling and Marketing: Strawberries are highly perishable and hence a great deal of care in harvesting and handling as well as its marketing also requires to be organised carefully. Usually the fruit is picked in the early morning and sent to the market in the afternoon of the same day or is picked in the late afternoon, stored overnight in a cool place, and sent to market the following morning.

Temperate Fruits

Apple (Pyrus malus)

Area and Production : Apple occupies nearly 12,141 hectares, mostly in the temperate regions of Punjab, Uttar Pradesh, Himachal Pradesh and Kashmir and to a small extent in the Nilgiris.

Varieties : Apple varieties fall into two categories; diploids and triploids. Diploids have plenty of good pollen and are self-fruitful. Triploids are self-unfruitful and become productive only when pollinated by using suitable pollenizer varieties. Even self-fruitful varieties have to be interplanted to get commercial crops through cross-pollination. Varieties selected for interplanting should sufficiently overlap in their blossoming periods. Important varieties are listed below;

Himachal Pradesh : Red Delicious, Golden Delicious, Worester Pearmain, Newton Wonder (all diploids) Cox's Orange Pippin (triploid), King of Pippins (No. 13), Starking (Royal) Delicious and Richard.

Kashmir Valley : Red Delicious (dipliod), Baldwin (triploids), Ambri Kashmiri, White Dotted Red and Blood Red.

Simla Hills : Beauty of Bath (triploid), Red Delicious, Jonathan, Rome Beauty (all diploids), Early Shanburry, Red Astrachan, Red Sudeley, Stayman Winesap, Winter Banana and Yellow Newton.

Kumaon Hills : James Grieve, Jonathan, Rome Beauty (all diploids), Blenheim Orange Pippin, Delicious, Early Shandburry, Golden Pippin, King of Pippins, Rymer and Winter Banana.

Kullu Valley : Ben Davis, Red Delicious, Golden Delicious (all diploids), Cox's Orange Tippin, Blenheim Orange, Baldwin (all triploids) Red Astrachan, King of Pippins, Yellow Newton and Granny Smith.

Nilgiri : Rome Beauty (diploid) and Irish Peach.

Propagation and Planting : Propagated mainly be shield budding, bench grafting and tongue grafting on seedlings raised from seed. Use M.IX dwarfing rootstock for propagating dwarf apple trees. One-year-old grafts are planted during the dormant period or in early spring 81.5 cm apart on M.IX dwarfing stock and 7.5 to 9 meters if on seedling stock. Standardized clonal rootstocks of the Malling Merton series are recommended where woolly aphis is serious.

Apricot (Prunus armeniaca L.)

Varieties : The cultivated varieties of apricot are mainly exotic and they grow successfully at varying elevations. The following varieties are recommended; Shipley Early, Kaisha, New Castle, St. Ambroise and Royal. All these varieties are self-fruitful.

Propagation and Planting : Apricot is propagated by shield budding on wild apricot stock, *i.e.* zardalu. Peach stock may also be used. Plant one-year-old grafts in autumn 6 to 7.5 metres apart.

Pruning : Apricot grafts usually have numerous lateral branches unlike the straight whips of apple, pear and cherry. If the laterals developed in the nursery are not properly spaced, cut off the main stem while planting, about 50 to 75cm above the ground level to promote the growth of new laterals. During the first summer, remove all unwanted laterals, leaving three to five well-placed branches to form the framework. Leave the upper branches longer than the lower ones, as the latter grow faster and crowd out the upper branches. By the end of the second growing season, the tree produces a large number of laterals on the seaffold branches and trunk, which should all be removed, except a few short growths (7cm to 12cm long) on the trunk and the main branches. Retain only five to seven secondary seaffolds. In subsequent years, thin only the branches, which are either crossing or crowding one another. This practice admits light into the centre and encourages the growth of spurs.

The pruning of old trees should aim at producing new spurs to replace those broken during picking. The kind and the amount of pruning depend upon the bearing habit of the variety. Light to moderate thinning of branches and the shortening of new wood back to the laterals is the usual practice. If new growth is less than 40 to 80cm each year, resort to severe pruning.

Manuring : A dose of fertilizers to supply 55 to 65kg of N, 55 to 65kg of P and 110 to 135kg of K per hectare may be applied to the bearing trees in spring.

Thinning of Fruits : Thin the fruits 4 to 8cm apart, leaving not more than two to three fruits on each spur.

Harvesting : The fruits should be picked when it is still hard, but has attained the proper colour. For drying, the fruit is harvested by hand-picking when it is fully ripe.

Cherry (Prunus avium)

Cherries are of Two Types: Sweet used for desert, and sour used for cooking, grown mainly in the Simla Hills, the Kulu Valley and Kashmir at elevations above 1,500 metres.

Varieties : Selected varieties of proven merit are;

Early Rivers, Governor Wood, Bigarreau de Schrecken, Elton, Bedford Prolific, White Bigarreau, Monstrue use de Mezel, Bigarreau Napoleon, Emperor Frencis and Late Black Bigarreau. It is desirable to choose varieties that will ripen in succession in order to obtain the crop over a longer period. A large number of varieties are self-unfruitful and do not set fruit with their own pollen. As they are also cross-incompatible, only the compatible varieties, whose period of flowering overlaps to effect cross-pollination, should be interplanted to get commercial crops.

Propagation and Planting : The plants are propagated by whip or tongue grafting on seedlings of wild cherry stock, called paja. Grafts are ready for transplanting in two years. Sometimes, the rootstock plants are planted in permanent positions in the orchard and grafted in situ.

As cherry-trees are generally affected by frost, the site selected for planting should be such that the sun reaches the trees gradually. The distance between the trees varies from 9 to 12 metres, depending upon the variety. The trees should be properly staked after planting.

Pruning : Cherry-trees grow into shape without much pruning. Crowded branches should be thinned out and dead-wood removed in the dormant season. The pruned cuts should be painted with tar.

Manuring : Cherry orchards are best put under grass, which is grazed by sheep. In addition to sheep manure, phosphate manures are applied to obtain a good growth of cloves. A dressing of fertilizers to supply 75 to 100 kg of N, 55 to 90 kg of P2O5 and 110 to 165kg of K2O per hectare may be recommended.

Peach

The area under peach (Prunus persica (L) Stockes) is very small and is mainly located in the Himalayas at various elevations.

Varieties : Some of the promising varieties are Early Beatrice, Alexander, Early Rivers, Duke of York, Peregrime, Noblesse, Late Devonian, Elberta, J.H. Hale and Triumph. Except J.H. Hale, all other varieties are self-fruitful and set good crops without cross-pollination.

Propagation and Planting : Propagation is done by budding on seedling peach. One-year-old grafts are planted 6 to 8 metres apart I early spring. Immediately after planting, the trees are white-washed to protect the bark from the sun.

Pruning : At the time of planting, the stem is cut to about 0.6metre from the ground and three to four branches are allowed to develop, distributed round the main stem. All other shoots that grow during the first summer are removed. During the first dormant season, two well-spaced secondary branches on each main branch are selected and the main branch is cut close to the secondary branches.

During the second summer, water-sprouts, if any are removed. At the time of second pruning in winter, secondary branches are not cut, except to regulate the shape of the tree. In pruning, cut always to the outside buds to encourage a spreading shape. In the case of bearing trees, annual pruning is necessary to maintain the centre open. Two to three-year-old branches may be cut back to the outward-pointing side branches to encourage a spreading growth. Shorten and thin outside branches to stimulate the growth of new fruiting wood every year. A satisfactory annual growth should be 45 to 50cm long. Fruit-buds are borne laterally upon one-year-old wood and on short spur-like twigs. Ordinary, they develop two fruit-buds and a leaf-bud at one node. The fruit buds are usually located from the middle of the shoot upwards. In cutting away branches, the position of the fruit-buds should be taken into consideration.

Culture : A peach orchard should be regularly cultivated. Ploughing, which should not be deeper than 10cm, is generally done in winter. A suitable cover or green-manure crop may be sown in the rainy season after the fruits are picked and ploughed-under during winter. A dose of fertilizers to supply 55 to 65kg of N, 55 to 65kg of P and 110 to 135kg of K per hectare may be applied to the bearing trees in spring. Immediately after the natural fruit-drop in May and June, the fruits should be thinned out so as to have them 10 to 15cm apart.

Harvesting and Marketing : Peaches are picked when they are still hard, as they can ripen well during storage or in transit.

Pear (Pyrus communis L.)

Pear is grown mainly in the hills at elevations ranging from 1,500 to 2,500 metres. Its cultivation is rather restricted, mainly because the fruit does not store well.

Varieties : The following varieties are recommended;

Willan Bon Christien (Bartlett), Clapp's Favourits, Thimpsons, Doyenne du Comice, Easter Beurre, Winter Nalis, Conference, Dr. Jules Guyot, Marie Louise d'Uccle, Baggugosha (Citron des carmes) and Emile d'Heyst. Baggugosha can also be grown in the submontane tracts, but there its quality is poor. Nashpati is another variety that is grown successfully in the plains.

Pollination : Most of the pear varieties are self-unfruitful and the planting of pollenizer varieties is advocated. Nashpati is a self-fruitful variety.

Propagation and Planting : Pear is propagated by shield-budding which is done in June-July. The stocks are raised either from the seeds of a commercial variety or from those of wild pear, shegal (Pyruspashia). To produce dwarf trees, quince 'C' stock is employed. Some varieties are not compatible with quince. They are propagated by double-working, using as intermediate a pear variety which can successfully be grown on the quince stock. One-year-old grafts are planted in autumn. Those propagated on the quince stock are planted one to one and a half metres apart, if they are trained as cordons. Those trained as pyramids are planted 3.5 to 4.5m apart. Grafts propagated on semi-dwarf stocks are trained either as bushes or modified leaders are planted 4.5 to 6m apart.

Pruning and Training : Pears on the pear stock make vigorous growth as develop into large trees. They remain dwarf on the quince stock when they are trained into different forms. For the pyramid form, cut the graft while planting at about three-fourths of a metre above the ground. Next winter, prune the leader to about 25cm and the laterals to about 20cm to the outward-pointing buds. In the second summer, all the branch leaders are laterals should be pruned to give to six leaves from the clusters, allowing the central leader to grow unchecked.

In the third winter, the central leader is cut back to about one-third its length, but the branch leaders and laterals are not pruned. In the third summer, the branch leaders and the laterals, except the central leader, are again cut back to five or six leaves as in the previous summer. In the fourth winter, the central leader is again cut back to one third its length. By following this procedure, a pear-tree on the quince stock would start flowering in fourth year. The bearing pear-trees are pruned as in the case of apples.

Fruit Thinning : As a rule, less thinning is required in the case of pears than in the case of apples. One fruit per cluster in the case of prolific varieties and one to two fruits per cluster in other may be retained after thinning.

Manuring : The method of manuring and the time of its application are the same as for apple. The amount of nitrogen to be applied should, however, be a little more than in the case of apples.

Harvesting and Marketing : Bartlett is picked when still green and hard. The early varieties are packed without storing, whereas the late varieties require storing to develop full flavour. The fruit should be size-graded before storing.

Strawberry (Fragaria spp)

The cultivated varieties of strawberry grown in India are all imported. The following are recommended; Laxton's Latest, Royal Sovereign, Early Cambridge, Huxley Giant, Phenomenol and Robinson.

Propagation and Planting : Maiden plants (runners) that have not borne any crop are used for planting. The planting-distance is half a metre between plants and three fourths to one metre between rows. Runners with a good root-system are used to set a new plantation. Transplanting is done in March-April in the hills and in January-February in the plains.

Culture : Prepare the land by ploughing deep, followed by harrowing. Add bulky organic manures. Keep down weeds by light hoeing and runners, as and when they form. Manuring is done in winter. When plants blossom in spring, bed the plantation with straw to keep fruits off the soil. After fruiting, remove the straw and weeds and cut off all runners. Continue hoeing. Rotate strawberry with vegetables every three years.

Irrigation : Apply irrigation at five-day intervals during summer.

Persimmon (Diospyros kaki)

Persimmon is grown in the Kulu Valley at elevations ranging from 900 to 1,500 kilometres. The promising varieties are Fuy, Hachiya, and Hyakume. Several good varieties, such as Dai Dai, Maru and Tenanshi, are also being grown successfully at the pomological station, conoor in the Nilgiris in southern India. The tree is propagated by grafting (Whip and tongue method) on seedlings of Diospyros lotus and D. viriginiana. The grafts are planted in winter, 6.5 to 7.5m apart.

The trees are headed back one or two metres above the ground at the planting time. Four to five shoots are allowed to grow round the stem to avoid narrow crotches and to develop a well-balanced head. There is no further pruning after this. Dead, broken and interfering branches are removed every year. The fruit is picked when it has attained a yellowish or reddish colour, characteristic of the variety, when still hard. It is clipped from the tree, keeping intact the calyx and a short piece of the stem. It is wrapped up in tissue paper and packed in a two layer box for transport. With astringent varieties, the fruit has to be cured before it is fit for eating out of hand. The simplest method is to place the fruits in a closed chamber with other ripening fruits such as pears and tomatoes.

Plum (Prunus domestica)

Plum is grown mainly in the Himalayas where the following varieties have been successfully grown; Grand Duck, Early Transparent Gage, Victoria, Santa Rosa, Wickson, Beauty and Kelsey. Nedu state, several choice varieties of the Japanese plum (P. salicina) are grown. The more important of these are Rubio, Alu Bokhara, Gaviota, Shiro, Combination and Hale. All varieties, except Beauty, Santa Rosa, Gaviota, Rubio, Alu Bokhara and Hale, which are self fruitful, require cross-pollination from other varieties. Plums are usually propagated by shield-budding on wild apricot or common peach stock.

Planting, spacing, cultivation and fertilization are the same as for peach.

Pruning : Cut back the top to about 60cm at planting time. Select three to five scaffold branches situated spirally round the stem, equidistant from one another, and remove the unwanted ones. At the time of first winter pruning, the main branches are headed back. All growth, except the main and secondary branches, is removed during the year.

At the second winter pruning, crossing and other undesirable branches are removed. In the case of varieties having a tendency for upward growth, heading should be done to outward-pointing buds to make out every year on similar lines. The pruning should be light as far as possible. The bearing trees are pruned to secure a balance between vegetative growth and fruiting.

Thinning : Thinning should be carried out after the natural fruit-drop in April and May but before the hardening of the pits.

Harvesting and Marketing : For transporting, the fruits are picked a few days in advance of full maturity. The change of colour for each variety determines its stage of maturity. The fruits are required to be harvested in three or four pickings.

Bibliography

Adams H.: *Hot Pepper Improvement*, St. Michael, CARDI, 1997.

Agarwal P. K.: *Improvement of Citrus*, New Delhi, Malhotra Publishing House, 1993.

Alexander M. P. and Ganeshan S.: *Pollen Storage*, New Delhi, Malhotra Publishing House, 1993.

Andrews J.: *The Domesticated Capsicums*, University of Texas Press. 1995.

Andrews L.: *Citrus Production - Orange*, St. Augustine, Trinidad and Tobago, 1990

————— : *Production of Minor Crops Sapodilla, Tamarind, Guava and Genip*, Port of Spain, Trinidad and Tobago, 1996.

Arguello D. S.: *Production, Post-harvest Management and Exportation of Tropical Fruits in Costa Rica*, Wageningen, Technical Centre for Agricultural and Rural Cooperation, 1992.

Ashworth S.: *Seed to Seed*, Decorah, Seed Savers Publications, 1991.

Barbeau G.: *Tropical Fruits in Nicaragua*, Managua, Nicaragua Ministerio de Desarrollo Agropecuario, Agraria, 1990.

Batlle I. and Tous J.: *Carob Tree (Ceratonia siliqua L.)*. Rome, International Plant Genetic Resources Institute, 1997.

Berger, P.L.: *Pyramids of Sacrifice: Political Ethics and Social Change*, New York, Basic Books, 1974.

————— : *Pyramids of Sacrifice: Political Ethics and Social Change*, New York, Basic Books, 1974.

Brooks, D.: *Water: Local-Level Management*, Ottawa, International Development Research Centre, 2002.

Bunnik J. S. C.: *Fresh Fruits and Vegetable: a Survey on the Netherlands and other Major Markets in the European*

Community, Netherlands, Centre for the Promotion of Imports from Developing Countries, 1990.

Burton W.G.: *The Potato*, Holland, H. Veenman & Zonen N.V., 1966.

Cao Van P.: *An Integrated Approach for the Production and Processing of Minor Fruits in Martinique (FWI)*, Port of Spain, Trinidad and Tobago, 1996.

Cardona M. J. G., Carvajal S. L. B., Salinas D. G. C. and Isaza R. G. B. : *Proceedings of the International Seminar on Plantain Production, Quindio, Colombia,* Quindio, Corporation Colombiana de Investigation Agropecuaria, 1998.

Chadha K. L. and Pareek O. P.: *Advances in Horticulture: Fruit Crops,* New Delhi, Malhotra Publishing House, 1993.

Chambers, R.: *Rural Development: Putting the Last First*, London, Longman, 1983.

Collymore L.: *Fruit Production in Barbados*, Port of Spain, Trinidad and Tobago, 1996.

Coste R.: *Coffee: the Plant and the Product*, London, MacMillan, 1992.

Crucefix D.: *Avocado Variety Selection for Export Development,* Roseau, CARDI, 1996.

Cull Brian & Lindsay Pax: *Fruit Growing in Warm Climates for Commercial Growers & Home Gardeners*, Australia, Reed Books, 1995.

Currah L. and Proctor F. J.: *Onions in Tropical Regions*, Kent, Natural Resources Institute, 1990.

Daniells J.: *Illustrated Guide to the Identification of Banana varieties in the South Pacific*, Canberra, ACIAR, 1995.

Degras L.: *Yam: a Tropical Root Crop*, Wageningen, CTA/ MacMillan, 1993.

Dhatt A. S. and Singh Z.: *Propagation and Rootstocks of Citrus*, New Delhi, Malhotra Publishers, 1993.

Diederichsen A.: *Coriander (Coriandrum sativum L.).* Rome, International Plant Genetic Resources Institute, 1996.

Doijode S. D.: *Seed Germination in Fruits*, New Delhi, Malhotra Publishers, 1993.

Dudley, E.: *The Critical Villager: Beyond Community Participation*, London, Routledge, 1993.

Featherly H. I.: *Taxonomic Terminology of the Higher Plants*, USA, Iowa State College Press, 1954.

Ferentinos L.: *Proceeding of the Sustainable Taro Culture for the Pacific Conference*, Honolulu, HITAHR, 1993.

Forde S: *Proceedings of CARDI/CTA Workshop on Marketability of Caribbean Minor Fruits*, Port of Spain, Trinidad and Tobago, 1996.

Georges, S.: *The Debt Boomerang: How Third World Debt Harms Us All*, Boulder, Westview Press, 1992.

Glowinski Louis: *The Complete Book of Fruit Growing in Australia*, Australia, Lothian Publishing Company Pty. Ltd., 1991.

Godden G.: *Growing Citrus Trees*, Australia, Lothian Publishing Company Pvt. Ltd., 1988.

Gowen S.: *Banana and Plantains*, London, Chapman Hall, 1996.

Green S.K. and Kim J. S.: *Sources of Resistance to Viruses of Pepper (Capsicum spp.): a Catalogue*, Taipei, AVRDC, 1994.

Gunjate R. T. and Tawde A. B.: *Propagation and Rootstocks of Mango*, New Delhi, Malhotra, 1993.

Hartley W.: *A Checklist of Economic Plants in Australia*, Melbourne, C.S.I.R.O., 1979.

Harwood, R. R.: *Marshalling Technology for Development: Proceedings of a Symposium*, Washington, DC, National Academy Press, 1995.

Herklots G. A. C.: *Vegetables in South East Asia*, London, George Allen & Unwin Ltd., 1972.

Hessayon D. G. Dr.: *The Vegetable Expert*, England, PBI, Publications, 1985.

Huaman Z.: *Descriptors for Sweet Potato*, Rome, International Board for Plant Genetic Resources, 1991.

Hurst Jacqui & Rutherford Lyn.: *A Gourmet's Book of Mushrooms & Truffles*, Sydney, Golden Press Pvt. Ltd., 1991.

Jacquat Christiane: *Plants from the Markets of Thailand*, Bangkok, Duang Kamol, 1990.

Jeffers P.: *Evaluation of Four Onion Varieties in Montserrat*, Plymouth, CARDI, 1992.

Kenridge K. C. and Hardy B.: *Biology and Agronomy of Forage Arachis*, Cali, International Centre for Tropical Agriculture, 1994.

Kroll R.: *Cut Flowers*, Wageningen, CTA, 1995.

Kunelius T.: *Annual Ryegrasses in Atlantic Canada*, Ottawa, Agriculture Canada, 1991.

Lofgren, H., Richards, A.: *Food Security, Poverty, and Economic Policy in the Middle East and North Africa*, Washington, DC, International Food Policy Research Institute, 2003.

Mabberley D. J.: *The Plant-Book : a Portable Dictionary of the Vascular Plants*, Cambridge, Cambridge University Press, 1997.

Madulid Domingo A.: *A Pictorial Cyclopedia of Philippine Ornamental Plants*, Philippines, Makati Metro Manila, 1995.

Malins A.: *Postharvest Handling of Pineapple and Mango*, Port of Spain, Trinidad and Tobago, 1992.

Mannetje, L. T. & Jones, R. M.: *Plant Resources of South-East Asia*, Wageningen, Pudoc Scientific Publishers, 1992.

Matsuoka H.: *Cultivation of Panicum Genetic Resources for Evaluation of Characteristics*, Tokyo, JICA, 1997.

Mc Donald F.: *Plant Tissue Culture Manual for Yam, Cassava, Sweet Potato, Dasheen (taro) and Tannia (cocoyam)*, Roseau, Dominica and Cave Hill, 1993.

Mellor, J. W.: *The New Economics of Growth*, Ithaca, Cornell University Press, 1976.

Miller William: *Dictionary of English Plant Names*, London, John Murray, 1884.

Mitra S.: *Postharvest Physiology and Storage of Tropical and Subtropical Fruits*, Oxon, CABI, 1997.

Morris M. L. : *Maize Seed Industries in Developing Countries*, Colorado, Lynne Rienner, 1998.

Morton Julia F.: *Fruits of Warm Climates*, Miami, Julia F. Morton Publisher, 1987.

Murali T. P. and Duncan E. J.: *In Vitro Propagation of Banana through Aseptic Manipulation of Male Inflorescences*, Port of Spain, NIHERST, 1990.

Nabhan G. P.: *Wild Phaseolus Ecogeography in the Sierra Madre Occidental, Mexico: Areographic Techniques for Targeting and Conserving Species Diversity*, Rome, International Board for Plant Genetic Resources, 1990.

Nijdam J. & De Jong A.: *Elsevier's Dictionary of Horticulture in Nine Languages*, Elsevier Scientific Publishing Co. Amsterdam; New York.

Oka H. I.: *Origin of Cultivated Rice*, Elsevier, Japan Scientific Societies Press, 1988.

Oldham P.: *Cost of Production of Major Tree Crops in Dominica*, Roseau, Ministry of Agriculture, 1991.

Ou Yang Jue Ya: *Comparative Study of Mandarin and Cantonese*, Beijing, China Social Sciences Publishing House, 1993.

Percival John: *The Wheat Plant - A Monograph*, London, Duckworth & Co., 1921.

Pilgrim R.: *Post Harvest Handling of Minor Exotics*, St. George's, Grenada, 1996.

Ragone D.: *Breadfruit: Artocarpus Altilis (Parkinson) Fosberg*, Rome, International Plant Genetic Resources Institute, 1997.

Rehm Sigmund: *Multilingual Dictionary of Agronomic Plants*, Boston, Kluwer Academic Publishers, 1994.

Roy S. K.: *Research on Multipurpose Tree Species in Asia: In Vitro Clonal Propagation of Artocarpus Heterophyllus*, Bangkok, Winrock International Institute for Agricultural Development, 1991.

Singh H. P. and Chadha K. L.: *Genetic Resources of Citrus*, New Delhi, Malhotra Publishing House, 1993.

Sinha G. C., Reddy Y. T. N. and Singh G.: *Propagation and Rootstocks in Guava*, New Delhi, Malhotra Publishers, 1993.

Sperling L. and Berkowitz P. *Partners in selection: bean breeders and women bean experts in Rwanda.* Washington, USA: CGIAR, 1994.

Sreekumar V., Indrasenan G and Mammen G.: *Studies of the Quantitative and Qualitative Attributes of Ginger Cultivars*, Calicut, Kasaragod, 1990.

Stover R. H. and Simmonds N. W.: *Bananas*, United Kingdom: Longman Scientific and Technical, 1991.

Thavarasook C.: *Proceedings of the Mungbean Meeting 90, Chiang Mai, Thailand,* Bangkok, Tropical Agricultural Research Centre, 1991.

Thomas E.: *Fruit Production in St. Kitts and Nevis*, Port of Spain, IICA, 1996.

Vargas E. M., Macaya G., Baudoin J. P. and Rocha O. J.: *Variation in the Content of Phaseolin in Wild Populations of Lima Beans (Phaseolus lunatus L.) in the Central Valley of Costa Rica.* Plant Genetic Resources Newsletter, 2000.

Webb M.: *Weed Problems and their Control in Rice, Papaya, Citrus and Sugar cane: a Report to the Belize Plant Protection Service*, United Kingdom, Natural Resources Institute, 1993.

Whealy K.: *The Garden Seed Inventory*, Decorah, Seed Saver Publications, 1988.

Whitwell A.: *Dominica Orchard Crop Management and Research Project: Pest and Disease Management,* United Kingdom, Natural Resources Institute, 1991.

Woolfe Jennifer A.: *The Potato in the Human Diet*, Cambridge, Cambridge University Press, 1989.

Yoshida T.: *Cultivation of Citrus Genetic Resources for Evaluation of Characteristics*, Tokyo, JICA, 1996.

Index

A

B

C

M

N

O

P

R

S

T

U

V

W

❑❑❑